J. Richards

**On the Arrangement, Care and Operation of Wood-Working**

**Factories and Machinery**

J. Richards

**On the Arrangement, Care and Operation of Wood-Working Factories and Machinery**

ISBN/EAN: 9783337864927

Printed in Europe, USA, Canada, Australia, Japan

Cover: Foto ©berggeist007 / pixelio.de

More available books at **www.hansebooks.com**

# OPERATOR'S HANDBOOK

FOR

# WOOD-WORKING MACHINERY.

# OPERATOR'S HANDBOOK

FOR

# WOOD-WORKING MACHINERY.

ON THE

ARRANGEMENT, CARE, AND OPERATION

OF

# WOOD-WORKING FACTORIES AND MACHINERY:

FORMING

A COMPLETE OPERATOR'S HANDBOOK.

BY

J. RICHARDS.

E. & F. N. SPON, 125, STRAND, LONDON.

NEW YORK: 35, MURRAY STREET.

1885.

# NOTE.

The present work is a revised edition of 'The Operator's Handbook,' published in London 1873. The sale of the first edition was mainly in Europe. The revision includes considerable new matter.

J. R.

## PREFACE TO THE FIRST EDITION.

In the 'Treatise on the Construction and Operation of Wood-working Machines' it was necessary to introduce a large number of expensive engravings, and treat of many things not directly connected with the processes of wood conversion, but relating only to the construction of machines. This, while it added to the value of the work for engineers and machinists, at the same time increased its cost, and placed it beyond the means of wood-working mechanics generally; besides, the plan of the work did not include the practical operation and care of machines.

In view of this fact, it has been considered expedient to supplement the 'Treatise on the Construction and Operation of Wood-working Machines' with a shorter one, directed to their care and management, including the plans of arranging and equipping factories for wood work, and particularly the details with which the practical workman has to deal.

The work is mainly based upon American practice, which can hardly detract from its usefulness in other countries. The wood interest is more extended in America than elsewhere, and we have every reason to assume, that with the present facilities of intercourse, wood conversion, like other manufacturing processes, will become more uniform in all countries, as it progresses and improves.

J. R.

# INDEX.

# INTRODUCTION.

At the present day it may be fairly claimed that machines have supplanted hand labour in working wood.

Year by year improvements have gone on, until bench work and hand skill have become comparatively unimportant elements in wood manufacture; and, as Professor Willis remarked before the Society of Arts, 1852, "Nothing remains to be done by hand, but to put the component parts together." None, except those who have learned the business when machines were not used, can realise this change. You may tell the workman of to-day of going out through the snow to a board pile, selecting stuff, carrying it in, and after scraping off the snow in winter, or sweeping off the dust in summer, laying out the stuff with a chalk-line, and straight-edge, ripping out the job by hand, then dressing it up with a jack plane. You may tell him of mortising by hand, cutting tenons and shoulders with a backsaw, and he will look at you with an incredulous stare. No wonder; this sort of thing has passed away, and with it, we are happy to say, some of the hardest labour that ever was dignified with the name of mechanical. It was mechanical nevertheless, and called for the continual exercise of judgment and skill; from the cutting out to the cleaning off, it was a kind of race between brains and muscle. But now machines do the work, and the main business of workmen is to take care of, guide, and direct them. The muscular work is gone; the brain work remains. We cannot quite say that wood workmen's occupation is, like Othello's, gone, but it is greatly changed.

Machine operating is a trade—not an ordinary trade, but one of great intricacy, one that cannot be completely learnt even in a lifetime.

A man endowed with a strong natural capacity may, during a long and diversified experience, become a proficient and successful operator of wood machines, but the incessant changes and improvements that are going on in both machines and processes, together with the arduous nature of his work, are more than enough to take up his time and his abilities. He is not a mechanic with a trade in the usual sense; but is a mechanic of many trades. The duties discharged by a machine operator in America, would be and are in Europe divided up into half-a-dozen different callings; there are, for instance, the sawyer, the filer, the planer, the jig sawyer, finisher, and others, involving a division of labour which would be very far from producing the results attained in wood-working establishments in America, where the machine operator must be a bench workman, understand all wood-machine processes, must be a machinist, not only one that can chip and file, but must know the theory of constructing and repairing machines; he must be a millwright, not an old time " whittler " who could pare for a week on half-a-dozen wooden cogs of a crown wheel, but a millwright who can lay out shafting, calculate speeds, build wooden drums and supports, and do it in a rapid and thorough manner; in short, be proficient in the most difficult kind of millwright work. Thus the wood workman, in escaping the muscular part of his calling, has only added to the mental part; but he has at the same time the assurance that the change dignifies his business, and leads to better pay, which has at all times and in all places corresponded more to the mental than the physical part of man's labour.

## Text Books.

Nearly every mechanical trade has its "Handbook," "Manual," or "Guide," based upon the practice of skilled men, and containing rules founded on experience, which have been of great use in giving information to workmen. To argue the merit of such books is superfluous. In every country the advancement of mechanic art has been largely if not mainly indebted to the dissemination of technical literature of this kind. A book relating to any branch of industry is, or ought to be, but the experience of some person, given with opinions and rules deduced from that experience, and is more valuable than oral instruction because more carefully given, can be often referred to, and used by a greater number of people. There has been in time past, and there is still, too much of a feeling that books cannot deal directly with practice, and relate to theory only ; and further, that theory and practice are not only different elements in mechanics, but in a measure antagonistic and opposed to each other. The further we go back, the more we find of this spirit, which has grown out of a variety of reasons, among which we will name the imperfection or impracticable character of certain books prepared by those who were only versed in theory, and did not understand practice as well. These things are mentioned as operating against the good that class-text books may do ; but still the fact remains, that to such books we have been in the past indebted, and to them we must in the future look as a principal means of disseminating technical knowledge.

We have said that nearly all mechanical trades have been developed by, and have, their text-books. Can anyone tell why wood manufactures have had no such text-books ? or rather, why wood working by machinery has had no books of any kind ? This is the more remarkable in America, where the wood-working interest is so extensive, and where at least half a million of people are concerned in wood manufactures. So long as the fact is assured, the reason is not important, except as it may tend to mend the matter in future.

We may say, that as changes and improvements in machines have been so rapid, text-books could not do much good ; that the art had no scientific base admitting rules that could be of general application ; and that the operations were too diversified in different branches to be treated under a general head, with other excuses ; but the fact still remains, without a sufficient reason, that wood manufactures have been greatly neglected, and that much that might have been done has not been done. In future, if the art is to keep up and maintain its place as one of the most important among American manufactures, it must, like metal work, textile fabrics, engineering, and other interests, have a literature consisting of text-books for operators and manufacturers, and a general system to guide, in the arrangement of factories, the operation and care of machines and like matters.

As to how far a text-book, or rather a handbook, may be of general application in wood work is confessedly a question of difficulty, and this should be considered in any estimate placed upon what is written upon the subject ; but there is still this argument in favour of having it relate to wood work in general, that the whole tendency of shop manipulation is to a uniformity of processes and machines, and the more of the work there is performed by machines, the stronger the analogy between different branches ; and also, as machines approach nearer and nearer to a standard form of construction for the general purposes of planing, sawing, mortising, and so on, the more uniform will be these processes. In short, the machines used for such purposes as joinery, cabinet making, carriage making, are becoming similar, except as to strength and capacity, which is not to be wondered at when we reflect that the one general principle throughout is cutting with sharp edges.

Hoping to contribute something to such a desirable end, this little treatise has been prepared. It is based directly upon American practice, which is peculiar, and could not be aided by text-books arranged for, and with reference to practice, in older countries, where labour is cheaper and the skill less ; where hand labour maintains a more important place and will no doubt for a long time to come.

We conclude this Introduction by further reminding the reader that in most mechanical trades a handbook would relate to processes alone ; but for reasons already given, a book for machine operators in wood manufactures must be more than this, or else fail to be of much use. It must to some extent treat of the construction of machines, the arrangement of wood manufactories, the power to drive them, the handling of material, of all that the machine hand has to deal with. As his calling is a combination of trades, so must this book relate to a diversity of subjects.

There is but little fear of going outside of what an operator has to do and know, for it comprises nearly all that is carried on in wood-working shops except the accounts, and often includes a liberal share in that department. With this fact in view, we have but little fear of getting wide of the subject, and are quite confident that although we may discuss things that the Title would hardly reach, we shall not go beyond what either belongs to his business or is of interest to the operator of wood-working machinery.

# WOOD-WORKING MACHINERY.

WOOD-WORKING establishments in America are divided mainly into those directed to the preparation of builders' material, the manufacture of furniture, and carriage work.

The first comprehend planing mills, door, sash, and blind factories, and moulding mills.

The second, all classes of furniture making, including chairs and turned work generally, with musical instrument cases.

The third, carriage work for railways and road traffic, with agricultural implements, a class of work that is analogous and, as a rule, performed on the same kind of machines.

Outside these three general divisions there are turning shops, bending works, handle factories, tool factories, and similar establishments, in which the processes and machines are more or less special.

Wood manufacture, as a process unlike most others for the conversion of material, is confined to a single operation, that of cutting, the nature of which will be treated of under another head. The principles being nearly alike in the action of all the different wood machines, it follows that the shops are, or can be, very much on the same general plan for the several divisions of work which we have named. The machines and the

B

material are nearly the same for general woodwork ; and if we except the sawing of rough timber, of which it is not proposed to say anything in the present work, rules that will apply to a planing mill, or furniture factory, will not be far wrong for a carriage shop or a car shop.

An ordinary wood-working factory may ˙be a plain rectangular building, not less than 48 feet wide inside; long enough and high enough to accommodate the requirements of the business. The writer in his experience has found 50 feet an advantageous width, and would recommend it never exceeding 60 feet; for beyond this the added width will not afford facilities in the same ratio, and will increase the proportionate cost of a building. A width of 50 feet to 60 feet will allow for what we will term four lines of machine work, two on each side, and a tram or waggon road in the centre.

The diagram given, Fig. 1, will serve as an example of

REFERENCES TO FIG. 1.

1.—Office, 14 × 16 feet.
2.—Counting room, 16 × 16 feet.
3.—Storeroom for oil, tools, and supplies, 10 × 16 feet.
4.—Repairing and tool-dressing room.
5.—Boiler-shed.
6.—Firing room.
7.—Magazine for shavings.
8.—Steam chimney.
9.—Engine-room.
10.—Steam furnace.
11.—Stairway.
12.—Hoisting platform.
13.—Cutting-off and jobbing saw-bench.
14.—Jointing saw.
15.—Jobbing saw.
16.—Large flooring machine.
17.—Matching planers for jobbing.
18.—Large moulding machine.
19.—Small moulding machine.
20.—Slitting saw-bench.
21.—General surfacing planer.
22.—Splitting saw for siding.
23.—Re-sawing machine.
24.—Waggon passage or tramway.
25.—Grindstones for planer-knives and tools.
26.—Engine lathe for repairing.
27.—Forge fire.
28.—Vice bench for machine fitting.
29.—Saw-filing bench.
30.—Pumps.
31.—Main driving pulley.
32.—Engine.
a a.—Shafting.

this arrangement for a jobbing mill. The plan is not assumed as presenting anything new, but is recommended rather for the opposite reason, because it is not new or ingenious.

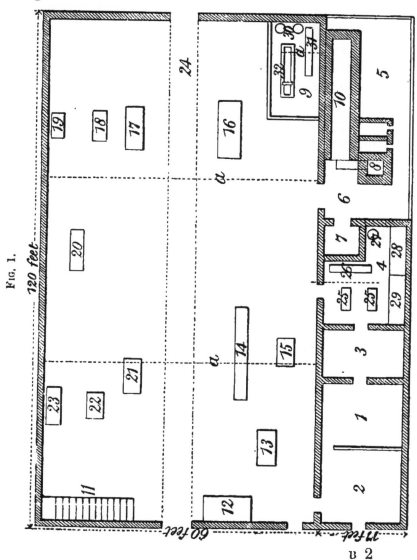

FIG. 1.

The most important matter to be guarded against in making plans for a new mill, is that of intricate and original designs, seemingly presenting great advantages on paper, and apparently quite correct to an architect or builder, but really quite wrong to a foreman or manager after a building is completed.

Fig. 1 is on a scale approximately as 1 to 200.

The plan here suggested is for a country jobbing mill 60 × 120 feet outside dimensions, having two cross lines of shafting, and equipped with machines requiring about 40-horse power.

The lower story should be 13 to 15 feet high in the clear, and the countershafts as far as possible overhead.

The arrangement of machines upon the floor is a matter that may be varied at pleasure, or to suit special kinds of work ; it cannot well be predicated upon an ideal plan, and can be remedied by changing, if wrong. The arrangement of the machines also depends upon their number and capacity. If in founding a mill the equipment is not complete, as is generally the case, there is no necessity for crowding and hampering machines to suit some general plan which may be carried out in future, when the mill is fully equipped ; it is often more advantageous to set machines temporarily, moving them as occasion may require, and thus obtaining more room, and greater convenience for the time being.

The shafting is shown arranged in two lines, but three are often better and more convenient. If three lines are used they will cost but little more than a single one running the other way of the building, and can have the advantage of being arranged to run at different speeds if required.

The last shaft, or the one farthest from the engine, can

be driven at a higher speed than the others to suit joiners' machines on an upper floor, an arrangement that is common in such mills; joiners' machines if driven from below will not require a line of shafting above, and a self-supporting roof can be used; the upper room may then be clear of posts, adding greatly to both the appearance and convenience of the room.

The position of the posts in the lower story is not marked in Fig. 1, but they can be arranged on each side of the central passage at a distance apart that will best accommodate the handling of long stuff, which is an important thing to be considered.

In connection with the plan Fig. 1, the following list of dimensions for machinery will be of use in making plans for mills, even when they may vary in capacity from the one assumed:—

Steam engine, 12 inches diameter, 20 to 24 inches stroke, with a speed of 75 revolutions a minute.

Boiler if double flued, 44 inches diameter, 28 feet long; if multiflued, one-fourth less capacity surface will do.

Grate surface, equal to 16 square feet.

Steam chimney, 60 feet high; area of flue, 500 square inches, fitted with air-tight slide damper.

Engine driving pulley, 10 feet diameter, 18 inches space.

Line shafting, 3 inches diameter throughout, to make 250 revolutions a minute.

Line-shaft pulleys, with average diameter of 36 inches and 12 inches face.

Average speed of countershafting, 750 revolutions a minute.

Hoisting platform, 10 × 6 feet.

As various dimensions will be hereafter considered under

separate heads, these are only given to render the diagram more complete.

For furniture and carriage manufacture, and in any case where the pieces are short, or reduced to short lengths, in working, the arrangement of machines must have reference rather to the course of the material through the shop as it is sawed, planed, bored, and mortised, than to providing room to handle it in.

In the case of a planing mill, a large share of the material worked is only dressed, or jointed and matched, and then again sent out; the trouble is to find room for it among the machines, and to handle it; in other words, to get it into and out of the mill without interfering with other work. If flooring is regularly or continually made, or if surfacing is continually going on, it is useless to provide room within the main building for storing either the rough or finished stuff; it should be fed in through the walls, and passed out of them as fast as worked, in such a manner as will not interfere with other operations going on at the same time.

A planing mill, where nothing but planing is done, requires a totally different arrangement from a mill where joiners' stuff and mouldings are made, or jobbing done.

The main building should be in such cases only 24 to 30 feet wide, with the machines placed side by side across the building, and have large doors opening opposite the feed end of each machine, as in Fig. 2.

The Figure is arranged on a scale of 1 to 200.

This plan in substance has been adopted in some of the larger mills in Chicago, and has many advantages to recommend it for a mill that is devoted to timber dressing alone.

It affords a mill of great capacity with but a limited

investment in the building, and the most economical arrangement of shafting and belts; besides, the plan is as safe from fire as it is possible to arrange one. The material is handled mainly out of doors, which gives

FIG. 2.

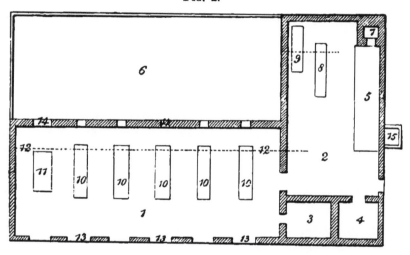

REFERENCES.

1.—The main planing room.
2.—The engine-room.
3.—Storeroom for oil, tools and stores.
4.—Magazine for shavings.
5.—Boiler furnace.
6.—Storing shed for worked timber.
7.—Steam chimney.
8.—Engine.
9.—Main driving pulley.
10.—Planing and matching machines.
11.—Surfacing machine.
12.—Line shaft.
13.—Large doors hinged at the top to open inward.
14.—Portholes for planed stuff to pass through.
15.—Ash-pit to the steam furnace.

unlimited room for storing, loading, and unloading it from wagons or railway trains.

The main mill-room and the engine-room should be thoroughly fireproof, with iron roof, and roof supports.

The walls should be 18 inches thick, and the overhead cross-beams not less than 15 feet above the floor, with the line shafting placed in pedestals, resting on top of the beams.

The line shafting should be 3 inches diameter, and make 250 revolutions a minute.

A mill of this capacity should manufacture at least 25,000 feet of matched stuff in a day, besides doing an equal amount of rough surfacing.

For general wood manufacture other than timber dressing or car building, the plain rectangular form of building represented in Fig. 1 is as nearly correct as any that can be devised. The material and the machines are short, and a given amount of floor room, with convenient ingress and egress, is all that is required.

Upper floors are, with good hoisting apparatus, nearly as good as ground floors for most purposes, and the most economical buildings for furniture manufacturing are from four to six stories high.

To secure good lighting, cheap timber framing, and to avoid posts, wood-working buildings should be narrow and long; or rather the width should be constant, and additional room secured by length.

A building for wood manufacturing can be made 48 feet wide in the clear, with a single row of posts in the middle, if the girders are deep enough, 16 × 12 inches for example, or if smaller they may be trussed, as shown in Fig. 3.

The truss rods are generally in the way of the bands, especially when the line shafting is placed, as it should be, across the building; and in nearly all cases it is both better and cheaper to provide strength in the girders without trussing them.

In the common plan of resting joists on the top, the girders are themselves in the way of the bands, and often cause great inconvenience.

But few ever consider in building shops that this method of mounting joists *adds their depth to the height of the walls;* so that it is not only an inconvenient but a very

FIG. 3.

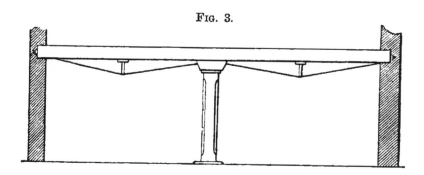

expensive one. A building with three floors will require to be some three feet higher at least, to give the same clearance between the floors as when the joists are let in flush.

For factories where there is overhead shafting, the joist should be gained into the girders, and rest on string pieces also, as in Fig. 4.

With bearing strips to help support the joists, the latter need not be gained into the girder deep enough to weaken it. The bottom of a beam is its weakest part, in resisting transverse strain; and a gain or notch say 2½ inches long and 6 inches deep in a girder 16 × 12 inches, does not affect its strength. The top receives only compressive strain, and after notching is generally stronger than the bottom side.

In Fig. 4, 5 5 are hanger-plates, which are thick enough

to come flush with the bottom of the girders, as shown by
the dotted lines. This arrangement of having the girders
project below the joist to a depth equal to a 3 or 4-inch
hanger-plate, is one that will find favour with any mechanic
who has had experience in erecting shafting beneath a
floor, where the joist was laid on the top of the girders,

FIG. 4.

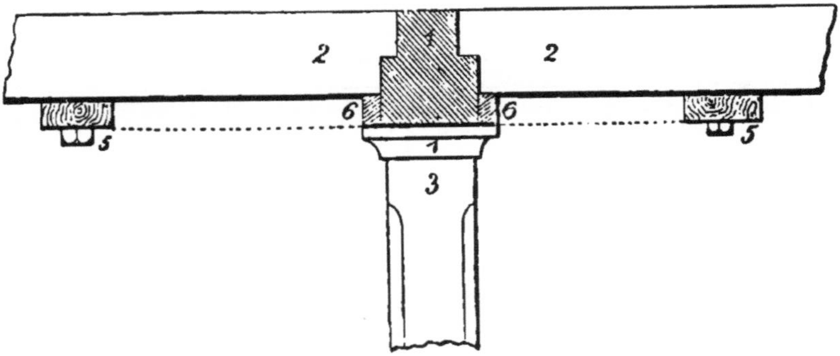

REFERENCES.

1.—Section across the girder.
2.—Joists.
3.—Post.
4.—Iron post cap, wide enough to receive the pieces 6, 6, which
     are bolted or spiked to the sides of the girder 1, to receive
     part of the strain and support the joists.

and where all the plans for bands, and even the position
of machines, had to be governed by the position of the
girders. As here arranged, the whole ceiling is in effect
a plane; a shaft or other overhead work can be set
anywhere. If a hanger comes on the girders, it is evident
that no hanger-plate is required, so that there is really no
inconvenience, but a decided advantage, in having the
girders project below the joist, to the difference of their
depth, say from 3 to 4 inches.

Joist floors are the best floors for wood-manufacturing establishments of all kinds. A plank floor, resting on girder beams, is very strong in the sense of supporting a great load, and will do very well for machine shops, but is totally unfit to resist the jar and vibration of high-speed machines. A floor of this kind is elastic and springy, no matter how thick it may be, while a joist floor, well bridged, is stiff and unyielding; although it might be broken through in spots with heavy weights, or might yield more in supporting heavy weights.

To put the same planking upon joists, that is usually laid on beams, would make a stronger floor in nine cases out of ten. But the custom is to put a thin floor, generally a single one, on joists, and a double one, consisting of heavy plank for the first course with $1\frac{1}{4}$-inch matched boards, upon beams. Without questioning the necessity of the double floor in the case of beams, and admitting that a joist floor is strong enough without it, it is certainly but fair to assume a floor of equal strength in the two cases, when making comparisons between the two methods.

A double floor is always best. One of $1\frac{1}{2}$-inch thickness laid across the joist, and an inch matched floor lengthwise of the building, making $2\frac{1}{2}$ inches in all, is strong enough for ordinary upper floors that have finishing machines to support.

Ground floors on which the heavy traffic comes cannot be made too strong. The weight of heavy machines requires good foundation supports to keep them level and to prevent vibration, but the piling of timber, which is quite as heavy, and falls first in one place and then another, is the main thing to provide against. The weight of machines is constant at one place, and when

once levelled up would remain so ; but if two to five
thousand feet of hard wood timber is piled near, unless
the floor is very strong, the machines will be listed over
or twisted by depression of the floor.

When there is no basement room, and nothing to hinder
building piers beneath a floor, there is no excuse for
having it weak enough to yield, and it only requires
proper precautions at the time of erecting the building.

STEAM POWER FOR WOOD-WORKING ESTABLISHMENTS.

Among other subjects which a foreman or wood-machine
operator is expected to understand is that of steam power.
The steam power is an integral part of the machinery of
such an establishment, and should not be conducted as a
kind of separate department from the rest. If it is, as a
natural consequence delays and derangements will be of
frequent occurrence.

To keep an engine always running requires quick
judgment and a fertility of expedients not often found
with the class of engine-men commonly employed in
wood-working mills.

In the United States foremen and operators are, as
a rule, well acquainted with steam power, and it often
becomes a part of their duty to give suggestions and
make plans for furnaces, boilers, engines, and other details
of the power department.

It is therefore considered quite in place to devote a
short chapter to the subject, directed to some of the
peculiar points to be observed in making plans for steam
power in wood-working establishments.

A wood-working factory, unlike a machine shop, has not the same facilities for repairing, and keeping fancy steam engines in order. The dust renders it almost impossible to keep them clean or bright, and the work is so irregular, and so heavy, that the expense of finishing is much better expended in more careful fitting.

The duty of a steam engine is not only more severe, but is more irregular than in almost any other business. As a rule, steam engines in wood-working establishments will be found working up to their full capacity, and require the packing and joints to be carefully kept in order. The duty is irregular in consequence of the sudden strain of starting planing machines, saws, and similar machines. The average duty is regular enough to dispense with independent cut-off valves on the engine, which must always add to the complication, and liability to derangement and wear. A strong plain engine is what is required, without bright finish or ornament, but with well-fitted joints and large bearing surfaces made of the best material.

The piston, cross-head, connecting rod, and main bearings are the vital parts to be looked after. The cross-head slides are continually deprived of their oil by fine dust that will find its way to the engine-room, no matter what precautions are taken; these should have either fibrous packing, oil feeders, or be made of wood. Strips of lignum vitæ will be found to wear well and be safe from cutting the slides; besides, they can be replaced at any time without detention.

An engine to drive wood machines requires a heavy balance wheel to ensure steady motion, it should have not less than 500 pounds of weight to each inch of diameter of the cylinder, and be as large in diameter as practicable.

The piston speed should for the same object be from 300 feet to 400 feet a minute.

The boiler and steam furnaces are matters of greater importance than the engine. They generate the power, the engine merely transmits it to the work, a thing not always thought of.

In determining which type of boiler to use, there are two leading conditions to be taken into account—the kind of water, and the kind of fuel to be used.

Wood refuse alone is not a strong fuel, but when mixed with bituminous coal it makes a very hot fire, which from its intensity and irregularity may be considered destructive to a boiler; to obviate this the boiler must be kept clean, and should be made of simple form, admitting of easy access to every part.

With hard lime water, which is commonly found throughout the middle States, this last-named condition becomes a necessity; no complicated multiflued or firebox boiler can last long when there is much lime in the feed water. The advantage gained by the thinner metal in the tubes or by the fire-box is soon lost through incrustation, while the original cost, subsequent repairs, cost of cleaning, care, management, and risk, are all in favour of the plain cylinder boiler without flues, or with flues that can be reached for the purpose of cleaning, both internally and externally.

The irregularity of firing with wood fuel, especially when a regulating damper is not used, makes steam room desirable; this is seldom obtained in a multiflued boiler, where the contracted heating surface generally leads to a proportionately contracted steam space, and this, with the ordinary mode of firing, has the steam " up " and " down " continually, causing a derangement of the work, and having a most destructive effect upon the boiler itself from intermittent strain. The heating surface and steam room, or in other words the capacity of a boiler, should be one-third more for a wood manufactory where the cuttings and shavings are burned, than where coal is exclusively used for fuel.

Although in opposition to popular opinion, a plain cylinder boiler without flues of any kind, carefully set in a first-class furnace, and made long enough to gain the full effect of the fire, is quite as good as any other. There is, however, not much use in recommending a thing which it is known will not be applied. There is a prejudice against cylinder boilers throughout most parts of the country that prevents their use in a great many cases where they would give as good a result as those with flues, and have other advantages which all must admit.

Following the general practice of the middle and western States, we present some views respecting the construction of furnaces for double-flued cylinder boilers.

The plans set forth in the Diagrams which follow have, for general objects, a tight furnace, a cool place to fire, and a saving in first cost, with greater safety from fire. Such a furnace as is here represented requires better mason-work than ordinary furnaces, and should have a thorough lining of fire-brick about the fire-bed. The whole amount of brickwork is greater than when an iron fire-front is used. As a modification of steam furnaces it may be considered adapted to wood-manufacturing establishments, because of its safety from fire and the avoidance of heat by the fireman ; the latter, considering the attention and time that is needed to fire with shavings, is no small object.

Fig. 5 shows a longitudinal section through a furnace

built with its end opposite to and combined with the
chimney, so that no breeching is needed. The firing is
effected from the side, as seen in the side
elevation, Fig. 6, without exposure to the
heat, and with more safety from danger
of fire. The ash-pit opens on the opposite
side of the furnace generally, outside the
building, where there is no danger of the
shavings catching fire while feeding the
furnace or when the attendant is absent.
A slide damper and the lever to work it
are shown on the front of the chimney,
Fig. 6.

Fig. 5.

A cross-section through the furnace at
the bridge wall is shown at Fig. 7, with
the covering over the boiler to retain the
heat and to guard against danger from
sparks. The filling, or covering, should
be of sand, earth, or ashes, instead of
mortar and brick, which is liable to crack
and allow sparks to escape when the
damper is shut; this is one of the most
common sources of fire about wood factories where steam power is employed.

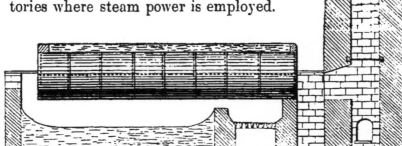

The following dimensions are for a furnace of this kind,

arranged for about 40 horse-power, and sufficient to drive a mill such as shown in Fig. 1.

FIG. 6.

Boiler, 44 inches diameter, 28 feet long.
Two flues, 16 inches diameter.
Height of steam chimney, 60 to 75 feet.
Area of flue in the chimney, 500 inches.
Area of boiler flues, 400 inches.
Area of throat at the bridge wall, 400 to 450 inches.
Area of grate surface, 16 square feet.
Area of the flue behind the bridge wall, 7 to 10 feet.
Clearance on the sides of the boiler, 4½ inches.
Clearance at back end of the boiler, 14 inches.
Size of fire-door, 15 × 30 inches.
Depth of ash-pit, 24 inches.
Width of ash-pit, 42 inches.
Ash-door (air inlet), 700 to 800 inches.
Thickness of furnace walls, single, 13 inches.
Thickness of furnace walls, if double, 17 inches.
Depth from boiler to grate, 18 to 22 inches.
Clearance between boiler and chimney, 24 inches.
The fire-room floor to be level with the grates.

A covering of loose earth or sand, as shown in Fig. 7, has other advantages besides the safety which it ensures from fire; it is cheap, easy to remove and renew, and a good non-conductor of heat. With a tight furnace covered in this

manner, it is comparatively safe to erect drying rooms

c

over a boiler, if the wood is kept at some distance—5 feet
or more—above the furnace.

The usual method of firing with wood shavings is wrong;
there are seldom any means employed to regulate the fire
or the quantity of steam generated,
except by the amount of fuel that is
fed to the furnace; a custom not only
wrong, because of the waste of fuel it
occasions, but because of the irregularity
it causes in the pressure of the steam
and the increased amount of labour in
firing. Without some means of con-
trolling the fire there is, at intervals, an
intense heat which generates more steam than is needed;
the fuel is soon burnt out, and the cold air allowed to pass
through the bare grates, until the heating effect of the fire
is in part counteracted. When fresh fuel is added it at
once burns up, or, as is often the case with a strong
draught, nearly all the lighter shavings are drawn over
the bridge wall before they are burned. An experiment
for a single day in the use of a regulating damper will be
sufficient to convince any one of its advantages. The
furnace should be kept full of fuel, no matter what its
character, and the steam regulated by the draught, either
with a slide damper operated by the fireman, or what is
much better, with a steam damper that regulates the
draught without any attention.

There are perhaps no simple contrivances that save
so much labour and money, so uniformly perform their
functions satisfactorily, are so much neglected and so
little known, as steam damper regulators. No one who
uses them would think of doing without them, and but
few who do not have them know of their importance.

FIG. 7.

There is no case where steam dampers are not needed, but nowhere else are they so important as to regulate the fire in the steam furnaces of wood-working establishments, where the fuel is of a mixed and inflammable character and cannot be fed with sufficient regularity to keep the steam at a uniform pressure.

In arranging steam plant for wood manufactories provision should be made to guard against freezing in the winter. Carrying out and bringing in such bulky material as timber always makes a shop cold, especially in the lower story where the steam power is placed. Nothing is more annoying than to be " froze up." A little oversight in this way often leads to expensive delay, when a small outlay would have saved all if it had been expended in time.

Another very important matter in the arrangement of steam furnaces for wood mills is to have them convenient to fire. It is possible to provide against heat by neither using a smoke breeching, nor an iron fire front, but if the fireman has to stand and shovel shavings through a small door breast high, only half has been done that can be accomplished to render firing easy. The fire-doors should be level with the fire-room floor, so the shavings and sawdust may be shoved into the furnace with a large scraper. Fire-doors should be not less than 30 inches wide, doors well lined to keep them cool, and the whole floor in front of the furnace made of iron plates, so that the fuel may lie about the floor without danger of catching fire, and thus avoid the trouble of continually sweeping up, which would otherwise be necessary. There is not the least objection to arranging a furnace in this manner, in fact there is a decided gain in convenience of access to every part, except to the ash-pit, which is but a small matter.

### SHAFTING FOR WOOD SHOPS.

If any machine operator of long experience, or, for that matter, of short experience, were asked what occasioned the greater number of accidents in wood shops and what caused most delays, he would be sure to reply, "The line shafting."

For a shaft to break by crystallization from bending—to be torn loose by winding bands—to have pulleys or couplings come loose, is a common cause of detention and expense. The couplings are mentioned last, although if ranked as to the amount of detention and trouble they cause, they should have been named first; but whether it be couplings, pulleys, hangers, or shafting, the trouble is generally with the "main line."

If we go to a machinist who makes shafting, and inquire whether there is any special difficulty in the way of having it safe from derangement or accident, he will answer, "Certainly not."

Granting this, we have either a paradox, or very bad practice, and as a paradox is rare in mechanics, the latter is the safer conclusion.

Shafts for transmitting motion and power are the oldest of mechanical appliances, and should, as we would suppose, for this reason, be among the most perfect, but this is a claim to which they can by no means pretend. The great diversity of the plans for couplings, hangers, and bearings by different makers attests the fact that the manufacture of shafting is by no means a perfected art.*

There are but few places where line shafting is so severely

---

* It will be fair to say that a few leading makers have so improved fittings that their shafting is almost free from the defects pointed out, yet a large share is still made on the old method.

tried as in wood shops; the usually small diameter, with high speed, wide bands, and the heavy duty that it generally has to perform, are conditions more or less avoided in other manufacturing establishments.

Machines when suddenly started, offer a resistance in proportion to the power employed in driving them, and measured by this rule, there are but few machines in common use so heavy to start and causing so great a strain upon the shafting, as planing machines and circular saws. There are of course many that require as much power, but to include all conditions, such as the speed of the bands and the usual means of shifting them, with the sudden stopping which often occurs, there is hardly a parallel among manufacturing machinery. A large planing machine or saw that consumes eight to ten horsepower to drive it, will have the bands shifted instantly from the loose to the fast pulley, and the only reason the shafting does not give way is that such machines are generally but weakly driven, and the bands slip until the machine gets into motion; the same thing in effect occurs in over-feeding saws, so that the shafting is continually subjected to a succession of torsional strains, that will soon search out bad jobs in fitting couplings or other parts.

In preparing plans for a wood-working mill, the shafting should, for reasons already given, go across the building whenever practicable. By connecting from one line to the other at one side of the room the whole power is not transmitted through couplings, as in the case of one continuous shaft to drive all the machinery. The work is also divided more evenly throughout the several lines, and this does away with the supposed necessity of having the line shafting in sections of various diameters, which prevents the interchange of pulleys from one shaft to another, and often leads to expense and trouble.

The first section of shafting carrying the main driving pulley should have a diameter equal to one-fifth the width of the main driving band, and be supported at each side of the main pulley; to make a rule, this section should not be more than twenty diameters long between bearings.

Fig. 8 shows a good arrangement of line shafting for a mill 50 by 150 feet, with three cross lines of shafting.

FIG. 8.

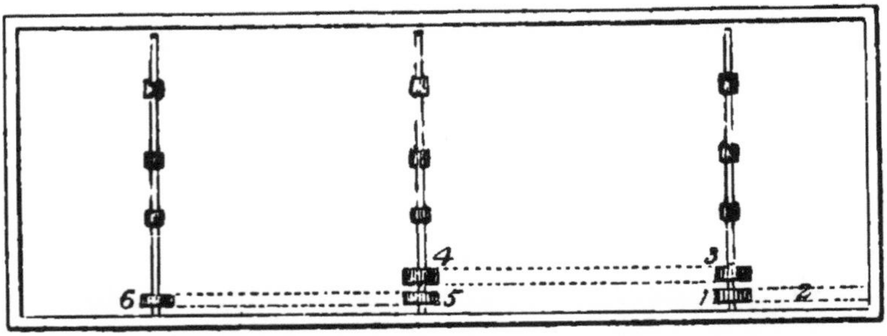

REFERENCES.

1.—The main driving pulley.
2.—Band to the engine.
3 and 4.—Second driving pulleys.
5 and 6.—Third driving pulleys.

Having the first or driving sections 6 feet long, and four additional sections in each line 10 feet long, is a good arrangement for a mill of the dimensions given.

The advantages gained by this plan over that of having a continuous line, or a single line running the other way of the building, are :—

First.—Only a part of the power is transmitted through the couplings.

Second.—The speed of the different lines can be varied

and to some extent accommodate machines of different classes, which can be arranged with this view.

Third.—A part of the shafting can be stopped for repairs, or to put on bands or pulleys without stopping the whole; in other words, about two-thirds of the works may be kept going in such cases.

Fourth.—With this arrangement the shafting can be of a uniform diameter throughout, except the first or driving sections.

Fifth.—The machines stand lengthwise the building, and the course of the stuff is in this direction, as it should be, and as it must be, for it is no uncommon thing to find planing and other machines driven with twist bands to accomplish this, when shafting is placed the other way.

For wood shops, 2½-inch and 3-inch shafting are the best sizes; 2½-inch shafts are as small as any should be, and they should not, without some important reason, exceed 3 inches in diameter.

A line of 2½-inch shafting will run safely and well at 300 revolutions a minute, or a 3-inch line will run 250 revolutions a minute, if the bearings are properly made and it is kept in line.

Pulleys should be turned true and balanced perfectly, no matter what their speed; it is never known where pulleys may have to be used, and the only safe rule is to have every pulley carefully balanced, no matter what the speed may be at which they run.

As to couplings, they should be adjustable or compressive, not keyed on, or "wedged" on as it may be called. Adjustable couplings are now very generally used for line shafting, and there is certainly no place where they are more required than in wood shops, where there is such a continual changing and adding of machines and pulleys, and the shafts constantly to be disconnected for the purpose.

Hangers to support the line shafting in wood shops should always have their bearings pivoted, and adjustable vertically. The heavy loads of lumber that are piled on upper floors depress them between the posts, and a line shaft requires to be often levelled up. If the bearings have a vertical adjustment in the hanger frames, and are moved by screws, as they should be, it is a small matter to go along the line and level it. A hundred feet of shafting may be adjusted in this manner in an hour, if the larger bands are thrown off to relieve it from strain, and the shafting is straight and true. The operation is so simple and so generally understood that it need not be explained here.

Shafting is not liable to get out of line horizontally, unless from the strain of bands ; it is, however, well to line up as often as twice a year, to be sure that all is right. It has been in times past a common thing to allow it to run as long as it would go, without adjusting, and then stop the works for a day or two to "line up"; which is unnecessary and only a loss of time. A shaft may be levelled by almost any one when the hangers are properly made, and may be done at noon, or after stopping in the evening, without interfering with working time.

To line a shaft horizontally is but little more trouble if the bearings or hangers can be moved in that direction.

Suspended hangers should have the bolt-holes slotted for an inch or more of movement, and post hangers should have movable bearings that permit side adjustment.

Assuming that there is some means of moving the shaft horizontally, a good plan of adjusting it is by suspending a number of plumb-lines that will bear against

one side of the shaft, and reach down low enough to be
sighted from the floor, as shown in Figs. 9, 10 ; or for

FIG. 9.

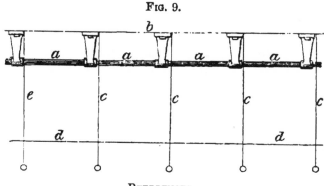

REFERENCES.

*b.*—The ceiling, to which the hangers are bolted.
*a a a.*—The line shaft.
*c c c.*—Plumb-lines resting against the shaft, near to the bearings.
*d d.*—A horizontal line stretched below the shaft.

greater accuracy a strong line may be stretched about 5 feet
from the floor, as at *d d,* to gauge the plumb-lines from.

This lower line can at the beginning be set within about
one-eighth of an inch of the two
plumb-lines at the ends, and the rest
can then be adjusted to the same
position by moving the bearings;
or the end bearings can be also adjusted, as the case may require.

FIG. 10.

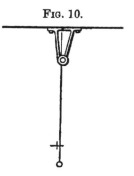

A ball of strong packing thread,
and half a dozen or more old screw
nuts for the plumb-lines, make an
outfit, and the job can be executed
with but little expense or time, if the hangers are properly
made.

This kind of work must be to a great extent a matter

of judgment; any one who depends wholly upon what he may have seen done and been instructed in will not be so successful in millwrighting as if he proceeds boldly, using his own judgment as to plans, and considering thoroughly the work before beginning it.

There are many ways of adjusting line shafting; some of them tedious and expensive. The one suggested is the most simple that can be given, and is accurate enough for all practical purposes.

## ERECTING COUNTERSHAFTING.

If a machine operator or even a regular millwright were to be set at a job to test his judgment and abilities, there is perhaps none that could be selected better than erecting a countershaft.

The ways of erecting, all of which may in the end produce the same result, are so various as to render it difficult to give rules that will be generally applicable. The advantages of the different plans can only be tested by the time required to do the work, assuming, of course, that it is to be properly done in all cases. It may require two, and often requires three, men a whole day to put up a countershaft, which in another case will be put up in two hours by one man, assisted only in holding and lifting.

In erecting a countershaft, first to be determined is the position of the machine to be driven, and whether the bands will be clear. When a line shaft is crowded with pulleys, it often requires great care to place the counter-shafts so that bands will not interfere with each other; it is no uncommon thing for a shaft to be put up, and then the discovery made that they interfere with others on the opposite side of a line shaft.

Care in starting is the main point, not only in putting up shafts, but in most other mechanical operations that involve calculations or accurate measurements.

Beginning with the hanger-plates, these should be of hard wood, long enough to reach from two to four joists, as the weight of the shaft and banding may require; their width should be from one and one-half to twice the width of the hanger base, and their thickness, as an approximate rule, one-fifth the drop of the hanger. When the joists are of hemlock, or harder wood, and three inches or more thick, almost any kind of shafting can be hung with safety on wood screws, or lag screws, as they are sometimes called, passing through the hanger-plate, and screwed directly into the joist. These screws should be of good size, not less than $\frac{5}{8}$ inch diameter in any case, and long enough to pass into the joist a distance at least equal to the thickness of the hanger-plate. A plate 3 inches thick requires, with cast-iron washers, screws that are 7 inches long; if one in each joist, $\frac{7}{8}$ inch diameter; if two in each joist, $\frac{3}{4}$ inch; or $\frac{5}{8}$ inch will do for ordinary countershafts.

Having the hanger-plates ready, next mount the shaft in the hangers and invert them on the floor, Fig. 11, and after settling the shaft to see that the bearings are not cramped, and that the hangers stand fair on their base, measure between the bolt holes accurately, or what is better, cut a short strip of wood to the length between the centres, marked $c$ in the figure.

If the shaft is to be placed to suit some pulley on the line shaft, measure from the centre of the hanger next the loose pulley the distance to the centre between the tight and loose pulleys; this should also be marked on the stick, as the *base* for the position of the shaft: we will term it the driving belt line, marked *a*, Fig. 11.

This band line must then be determined and scribed on

FIG. 11.

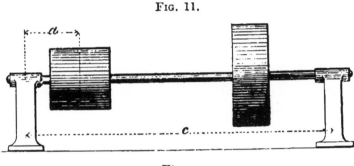

Floor.

the joist; it is easily found from a pulley, or by measuring from a wall or girder that crosses the line shaft at right angles.

Placing the measuring stick, next set out at each end for the wood screws or bolts that are to hold the hanger-plates, bore the hanger-plates and screw them up at one end, but not hard against the joist, leave a half-inch or more for packing, when levelling up; then set the plates at right angles across the joist, and mark the position of the joists so as to bore through the plates for the other screws, which can be done by swinging the plates around, and without taking them down. Again set the plates across the joist as accurately as possible by means of a carpenter's square, and mark the place for holes in the joist for the remaining wood screws. In screwing up the plates they can be brought level by furrowing down on their top, with pieces of wood split in two or notched to accommodate the wood screws.

To mount the hangers, if they have pivot bearings, as all ought to have, bore through the hanger-plate for one bolt by measurement; no great accuracy is required unless the shaft has to come laterally to a particular line, which is seldom the case. Screw up one hanger with a through

Fig. 12.

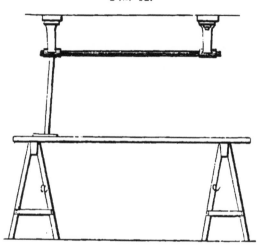

bolt, then remove the pulleys from the shaft, put it in the hangers, and propping the loose one, with a brace resting on the floor or a stage, as shown at Fig. 12. For the next operation, procure a pole or strip of wood c, Fig. 13,

Fig. 13.

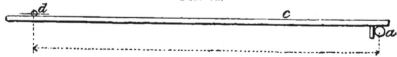

long enough to reach from the countershaft to the line shaft, cut a notch in the end, or drive a strong spike in the side, and let it rest on the line shaft, at a, and extend to the countershaft at d. By moving alternately from one end of the countershaft to the other, and driving the loose hanger to adjust it, a parallel is obtained, much truer than by lines and measurement, and in a tenth part of the time. The pole can be marked at the centres of the countershaft at each trial until the ends correspond. Then bore the remaining holes for the hanger-bolts, put the pulleys on the shaft, and mount the whole in place. Level the shaft by means of a spirit level or by a plumb-line alongside the pulleys, which, if they are at all true, will be found accurate enough. The work is now finished, and there is a question as to which is the greater labour, to erect a shaft or to describe the operation. With a good pair of trestles at hand, wood screws and hanger-plates ready, an ordinary countershaft for bands three to six inches wide should be put up in from one to two hours' time, by one man and an assistant. The time of erecting, and the accuracy with which a shaft can be set, as well as the facility with which it can be kept in line, depend greatly upon how the hangers are made. All bearings in wood-working establishments should be pivoted; the depression of floors which takes place is continually altering the bearings in a greater or less degree, and if rigid, they are spoiled by the least change. Such nicety is not required at low speeds, but when shafts carrying heavy strained bands have a speed of 750 or more revolutions a minute, every precaution must be observed to have them run without heating. If the bearings are pivoted, and arranged to be adjusted vertically, it is but little trouble to keep shafts level. The bolt holes in the hanger-plate if slotted to allow for horizontal adjustment, will answer for pendent hangers without having the bearings movable laterally in the brackets.

The transverse strength of the brackets should be sufficient to break the bands, if not, there is always danger of the whole being torn down by winding bands ; and as these are generally much stronger than are used in other shops the hangers should be made accordingly.

## SETTING MACHINES.

Setting machines belongs to the same class of work as erecting shafting, and is much the same thing and a matter of judgment rather than one of acquired skill.

The only general rule that can be given is to set them level, with their shafts and spindles parallel to the line shaft. There are, however, many plans of doing this and a word on the subject will not be amiss.

When a new shop is built, each floor should be scribed with what we will term a machine line, that is, a base from which the engine, the line shaft, countershafts, and machines may be set, independent of each other, and yet with accuracy. To do this, a centre line should be made through the building both ways, and scribed on the floors, not with an awl alone, but with a wagon maker's scribing hook, that will cut a deep groove. After striking with a chalk line, a straight edge should be fastened down and the lines scored so that they will remain as long as the floor lasts, or at least as long as machines are to be added.

If there are ground floors, the lines can be made on the walls, or ceiling ; they should be somewhere, in each story, and in each room. When these lines are once made, the setting of machines becomes a simple matter, lines parallel or at right angles are easy to lay out; and shafts or spindles can be set true by measurement as in Fig. 13, if they are first levelled.

A common practice when a shaft or machine is to be erected is to square it from something which has previously been set by something else, on the principle of measuring by succession, a practice no mechanic would think of in other cases.

If machines have iron frames and stand on masonry, they can be fixed by running melted lead or brimstone under the feet after setting and levelling them. On earth floors, however, it is not necessary to build masonry for any except reciprocating machines. Stakes of locust, cedar, or mulberry wood, set in the earth from three feet to four feet deep, and then sawn off level on top, make almost as good a foundation for any machine as masonry. It is, however, exceptional to find machines set on the ground, a plan that has nothing to recommend it, there has in any case to be a floor over a great part of the room, that usually costs as much as a complete floor would, if it had been laid down at the beginning.

---

### BANDS FOR WOOD MACHINERY.

Most rules that apply to bands in general are applicable to those used in wood-working establishments, yet there are some conditions to be taken into account that are peculiar and exceptional. They *dry* in all cases, and often have shavings or sawdust passing under the surfaces, preventing contact on the pulleys, and so reducing the tractive power; besides, the bands move at such a high speed that it prevents contact on the pulleys, especially when they are of small diameter.

For these reasons, bands should be much wider than

would be required to transmit an equal amount of power in other establishments.

Bands to drive wood-cutting machines require to be at least one-third wider than for metal-cutting or other machines where the bands can be kept soft or moist. Even twice the width will not be too much in some cases.

For main bands, india-rubber is preferable to leather. It has advantages in driving capacity, in running true, and, if well made, it is more durable ; its merits are, as a rule, not understood, although it has been in use for many years. The ordinary rubber bands of commerce may not be as durable as leather ; both the webbing and the rubber may be of poor quality ; but if an order is sent to a first-class firm for a good rubber band, heavy enough for its work, there is no leather band that will equal it. The driving power in a wood shop, where the surfaces must run dry, is at least one-third greater than that of leather, and the tension can be proportionately less, or the band proportionately narrower to do the same work. The best plan, however, is to keep the width and avoid tension, which, if too great, is apt to break the joints and heat the bearings of shafts.

For joining rubber bands there is no better plan than with malleable iron hooks. Clamps, with plates on the back, and other contrivances of a similar kind, make the joint too rigid, and also make a disagreeable noise in passing over iron pulleys. Cement joints that are generally recommended by the manufacturers cannot well be made by those unskilled in the matter, and are not necessary except for heavy driving bands.

What is wanted is a smooth joint, quickly and cheaply made, and one that will not pull out ; such a joint can be made with hooks. A band 12 inches wide can in this

D

way be put together in a good workmanlike manner in
ten minutes, and the joint will stand for a long time under
any strain that a band ought to bear, whether it be of
gum or leather.

To make the joint, cut the band square; then lay out

FIG. 14.

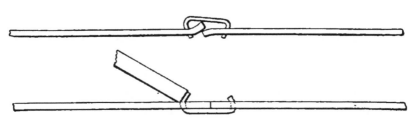

the lines for the holes, so that when the ends of the band
are placed together the distance between them will be a
little more than the length at *a*, Fig. 14. Punch the

FIG. 15.

FIG. 15*a*.

holes, then lap the ends, as in Fig. 15, and drive the
hooks by keeping a bar of iron, a hammer, or some other
weighty piece beneath the band. After the points of the
hooks are through at both ends, the join can be butted

together by bending the band backward from the joint until the ends will pass, and then straightening it. To clinch the hooks use an anvil bar, Fig. 15a, closing first one end and then the other with a light hammer, so that the band will be firmly clamped, *but not cut*, with the hooks. In this last operation lies the secret of making these hook joints successfully; if the hooks are closed properly they will not tear out the holes like lacing, but will pull the band asunder at the holes, proving the joint to be as strong as any other portion of the band, less the weakening effect of the holes. If the hooks are hammered down too hard they cut into the band and weaken it. After the joint is closed the hooks may be bent to conform to the curvature of the pulleys they run over. If one is large and the other small, the hooks should be bent to fit a curve between them in size, or to fit the small pulley.

That such hooks have not become more popular is owing to the careless manner in which they have been used. A band may be fastened in almost any manner with lacing, and hold for a time; but it is not so with hooks; they must be put in carefully. Properly done, they make one of the best joints, and if improperly done, perhaps the worst.

The size of the hooks must be adapted to the thickness and width of the bands; the distance from the joint to the holes should be at least equal to three thicknesses of the band.

The width of driving bands and their length should be such, that when at angles lower than 30 degrees they will do their work without tension on the slack side. By no tension, is meant that the band should be loose enough to hang in a curve. Main driving belts are here alluded to, and particular stress is laid on this matter, for no good result can be attained with a heavy band that is not capable of doing its work mainly by its weight.

Speaking of weight, it may be remarked that in making comparisons of cost between leather and india-rubber, weight should be taken into account. As a rule, single leather bands wider than 6 to 8 inches are not to be compared in weight to rubber ones of two and three ply, which with heavy cotton webbing, correspond to double leather bands, which are usually double the price. A leather band wider than 8 inches should always be double, no matter what its purpose, unless it is to run at a very high speed on small pulleys, which need never occur if machinery is properly arranged.

For the extreme high speeds sometimes necessary in wood machines, bands of cotton webbing can be used with advantage. Heavy saddlers' webbing coated with beeswax makes a band that is very light, and has a high tractive power. When used the pulleys must be true and smooth, and the bands kept clear of flanges, or anything that will produce a rubbing action, as this soon destroys them.

In the change from round bands, once almost exclusively used, to flat ones, we have no doubt gone too far; round ones are in many cases much cheaper and better. They are extensively used in England and on the Continent, but are rarely seen on American machines. For the first movers to drive the feed works of planing and other machines, they are better than flat bands, especially when cones are used for graduating the speed, and when they are exposed to shavings or sawdust.

In the treatment of bands for wood machines nearly all that can be done is to keep them soft; a coat of warm castor-oil now and then laid on with a brush or sponge is a good way to soften them. Tallow is as good, but more difficult to apply. For rubber belts no surface coating can be so good as the india-rubber itself, which is soluble in and infused by animal oils ; as such bands do not need softening they should be left alone, as the safest plan.

## HANDLING MATERIAL.

A large share of the labour of a wood-working establishment is directed to moving and handling material. It is one of those things which cannot be done to any extent by power; and in machine operations constitutes more than half the labour. There can be little information given about handling long timber, but the following suggestions in regard to short stuff or work in process will enable an operator to get along without so much handling and carrying as is common.

In arranging machines, they should always be set so as to leave truck-room between and around them ; no matter how crowded a room may be ; the floor-room saved by piling stuff on trucks will more than make up for that lost in passages.

In furniture and chair shops, carriage shops, turning shops, door, sash and blind shops, and in nearly all wood-working factories, the material can be kept on trucks instead of on the floor, and two important advantages gained ; it may at any time be moved from place to place, and can readily be reached without stooping to the floor.

We may also mention the system, order, saving from bruises, and the facility for counting pieces, as further objects gained by the truck system suggested.

The trucks for machine rooms should be made of uniform size for each story; there is no use in depending upon a particular truck being kept for a special use; the

FIG. 16.

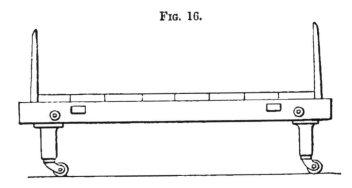

rule is, to take the first one at hand, and there is but little use in having different sizes. They can be made as shown in Figs. 16, 17, for stuff cut out and in process, and

FIG. 17.

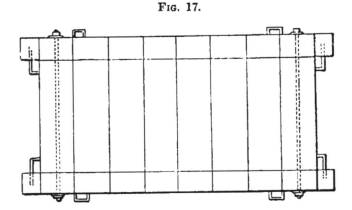

for anything except heavy loads of timber, which require a truck that is lower and much stronger. The main frames should be of hard wood, about 4 × 4 inches, the cross rails set in $3\frac{1}{2}$ inches from the end, with tenons to keep them in place. Two through bolts ⅝ in. diameter along the inside of the cross rails hold the frame firmly together, and yet allow it to yield in passing over blocks or uneven floors.

The common mistake in making such trucks is in having them too rigid; they will not last long or work well, unless made to yield at the corners. The planking across the top can be nailed to the side rails ; it should be 1¼ or 1½ inch thick, of white wood, sycamore, or some other tough wood, that will stand bruising, and not split; even pine is better than ash or oak. The standards should be arranged to go either at the ends or on the sides, as shown

FIG. 18.                FIG. 19.

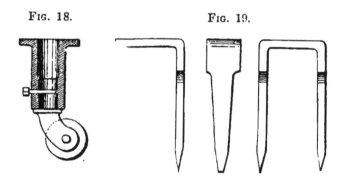

in the plan, Fig. 17. Figs. 18 and 19 show a complete set of irons for a truck 4 feet to 5 feet long and 2 feet to 3 feet wide, consisting of four cast-iron brackets with a flange at the top to be fastened with wood screws ; the swivel piece may be cast of malleable iron ; the small screw is to keep the swivel from falling out when the truck is lifted; the roller can be of cast iron ; the staples are for the sides and ends of the truck, as in Fig. 17 ; these staples should be forged from iron about 1½ × ¾ inch, and large enough to receive a tenon 2½ × 1½ inch.

With from six to twelve of these trucks on a floor, or at least one for each machine, half the handling and nearly all the carrying is saved. In working stuff two are needed at each machine, so that the pieces can be taken from one and placed on another as they are operated upon.

When material is to be moved from story to story, the trucks can be run upon the platform of a hoist, and with their loads raised or lowered to where they are wanted. A boy with one of these trucks will move a thousand pounds the length or width of the shop, and up or down through several stories, at the same cost that a single load can be carried by a porter, to say nothing of the damage by having the stuff thrown down upon the floor, and the loss of time required to gather it up again. This system of roller trucks is to some extent in use; but it is exceptional, and rarely ever carried out so as to realize the greatest advantage from it.

A system half carried out is as no system at all, and one or two trucks in a large shop are only an annoyance; the men lose more time during a year in searching or waiting for them and in disputing about them than a dozen additional new ones would cost.

To say that a wood-working establishment which has more than one story should have a power hoist, is to state what everyone knows, but not a thing which everyone has estimated the advantages of. A wood platform or cage, with a wire rope and winding drum driven by bands and a tangent wheel, is a cheap and simple plan for such hoists; the gearing is now furnished by different makers, like any other machines, self-contained and ready to erect, including the cage and guides if wanted. There should be a reliable safety catch to prevent falling; all ingenious triggers and self-acting apparatus can be dispensed with. A caution notice with directions for operating the machinery should be placed at each hatch, and the rest left to the judgment and good sense of workmen. There is no machinery so dangerous as that which pretends to dispense with care and caution on the part of attendants; and the greater number of accidents with hoists come from that class known as the "absolute safety." Accidents rarely happen with the old outside chain hoist, although it is without question very dangerous; the reason is that people watch it and run no risks.

In connection with the arrangement of a mill at Fig. 1, a tramway through the centre of the building is mentioned. This plan is a good one in a large mill or car shop, but in furniture factories, chair factories, door and sash shops, and jobbing mills, trucks such as those just described for machine rooms, only stronger, are more convenient than the tramway.

The general means of moving material may be said to consist in tramways for horizontal movement in straight lines, hoists for vertical movement, and caster trucks for distributing in irregular lines; however, in any but the largest mills, and for any but long and heavy timber, the horizontal movement and the distributing can be combined, and the fixed tramway dispensed with. In such cases the trucks to be used in connection with cutting out saws, planing machines, and for first floor purposes generally, should be framed of stuff about 5 × 5 inches, and be correspondingly heavy in all their parts; they should be from 6 to 8 feet long, with three wheels instead of four, the two forward wheels on a fixed axis, and the rear one swivelled. Such trucks should be strong enough to carry at least 2½ tons, and their wheels

8 to 12 inches diameter, with from 2½ to 3½ inches face. There is nothing peculiar about the construction that calls for diagrams to explain.

By laying a cheap plank floor from the mill room to, or through, a yard, such trucks can be run out and loaded at any distance from the shop, and men will prefer to push in a thousand feet of stuff in this way to carrying one or two boards.

This simple matter of trucks is dwelt upon because it is perhaps the most neglected of all things about wood shops. We exhaust our ingenuity in devising machines to work timber at a rapid rate, but make no provision to bring it to or from the machines; and with the exception of the large timber mills along the northwestern Lake coast, and the very largest mills in cities, it is unusual to find any means of handling material that at all compares with the completeness in other details.

### CLEARING WOOD SHOPS.

Clearing shops of cuttings, shavings and sawdust belongs to a certain extent to the same branch as moving and handling material, and the same rules will apply in many respects.

There is, however, this difference, that from recent improvements it is probable that the driving power will in future be used to clear shops, while we can hardly hope to have it handle material. Pneumatic conductors are now so well known that it will be unnecessary to go into a description of their general arrangement, which most readers are presumed to be familiar with. The writer having made pneumatic apparatus, which has been in constant operation since 1862, has no fears in recommending the system as practical and economical in most cases. Apart from its convenience, its sanitary advantages in getting rid of the fine dust is an important matter.

The fans should be plain, strong machines, large enough to perform their work easily ; the vanes strong enough to break up sticks that may be drawn in. The bearings should be *outside* the casing and pipes. A common plan is to have one bearing inside the induction pipe, where the oil is at once absorbed, and there is a continual danger of fire from the bearing heating. Fans made for ordinary blowing purposes are not suitable for conductors.

At Figs. 20 and 21 are shown side and front elevations of the fans designed by the writer for use in England.

The casing is in one piece ½ inch thick; the vanes are of forged or malleable iron ; the shaft is 1¾ inch diameter of steel running in brass bearings outside the casing.

The size of the fans for clearing wood shops must depend upon the number of inlets, openings, or, as we will call them, leaks into the induction pipes. An exhaust fan 20 inches diameter and 5-inch vanes, would clear the largest mill, so far as conducting the shavings and dust, but could not maintain a current strong enough, after supplying the inlets to lift shavings. For this reason it is easy to see the importance of having the collecting hoods fit well, and avoiding all possible leaks into the pipes. It is almost impossible to give any rule for the size of pipes without assuming some special premises to base such dimensions on. We will, however, say that starting with 5 inches diameter for the smallest size for a main pipe there should be added at least 10 inches of sectional area for each machine that is connected, except surfacing or dimension planing machines, which will need twice as much.

Galvanized or zinc-coated sheet iron, from 18 to 24 gauge, is a good material for conducting-pipes.

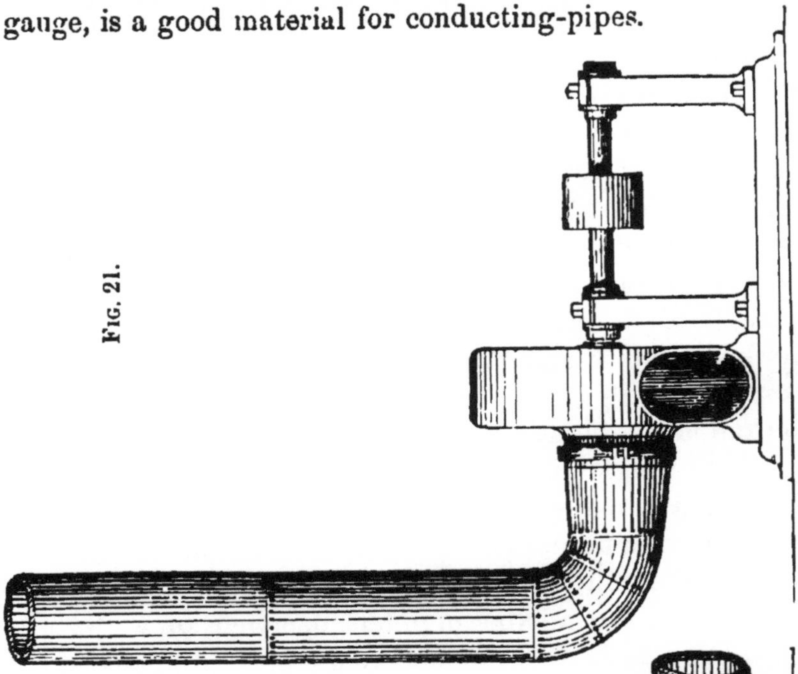

FIG. 21.

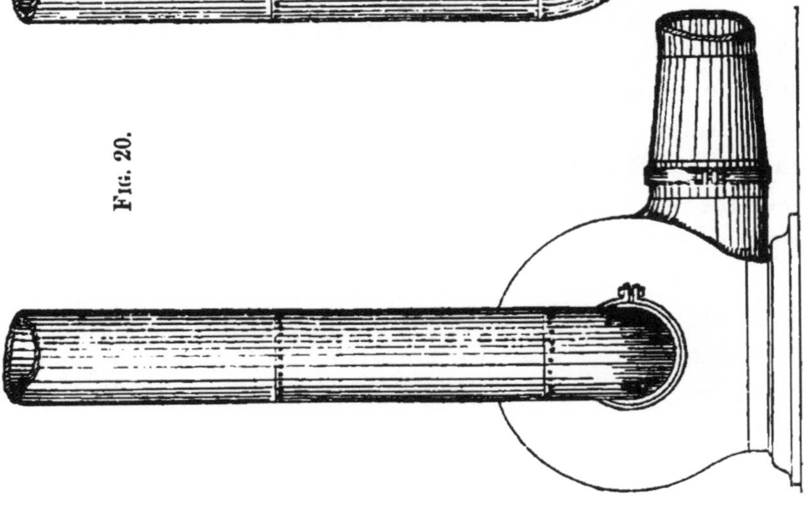

FIG. 20.

The elbows should be made with a radius of 10 inches or more on the short side, and everything avoided in the arrangement of the pipes that will endanger their choking. When machines are not in use, it is well to close off the induction pipes with a ball of paper or waste; dampers or valves can be made in the pipes for this purpose, but if constructed so that they will not obstruct the pipe when it is in use, they are expensive, and unnecessary, except for floor pipes, noticed farther on.

Fig. 22.

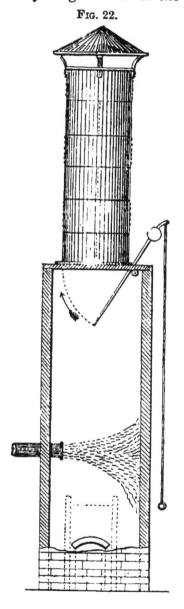

It is often desirable to have the fine dust separated from the shavings and sawdust; even if they are only to be used for fuel, and the magazine or shavings room should be arranged to allow the dust to pass off at the top, as in Fig. 22.

The magazines should be fireproof throughout, and extend above the building to such a height that the dust will not be carried through the windows after it has escaped at the top. As it is often expensive to carry the brickwork high enough to effect this object, a sheet-iron flue or up-

take can be used, as shown in Fig. 22. The sectional
area of this flue when used should be ten times as large
as the pipe leading into the magazine, otherwise the
current will be strong enough to not only carry off the
fine dust but the lighter shavings.

There should be a swing trap-door at the bottom of
the uptake, or at the top of the brickwork if an iron flue
is not used, that can be instantly closed from the outside if the shavings in the magazine should catch fire.
This trap can be pivoted on a shaft to extend out
through the brickwork, and be operated by a lever on
the outside.

The discharging door below should be closed by means
of a sliding iron plate, counter-weighted and working in
grooves, so that it will rest on the shavings when the
magazine is full, or partially full, prevent the dust from
escaping, and at the same time prevent any circulation of
air in the case of fire.

Inlets or openings, to take off sweepings, should be
provided at suitable places for clearing the floors. If
opening downward the orifices should be at least as small
as the pipe, and never made in a hopper form, as they
will soon be clogged with blocks or sticks.

A better plan for these floor openings for sweepings, is
to bring down a pipe from the main overhead, cutting
it away at one side, Fig. 23, and closing the aperture
with a slide door when not in use; this plan is much
better for many reasons than inlets cut through floors.
Such pipes can come down alongside a post or the wall
and not interfere with the room ; arranged in this way
there is but little danger of choking, or having lost tools,
nails, or blocks, drawn in. For conducting sawdust alone,
small tin pipes, 2 to 3 inches diameter, will do.

The danger of fire from such apparatus, once much apprehended, is owing to the use of wooden conducting pipes, having pockets and corners where fine dust would accumulate, and then explode by a spark communicated from a hot bearing, lucifer matches being dropped among the shavings, or by sparks from the fan striking grit or nails. The inflammable and explosive nature of wood dust is but little known; few are aware that it will explode like gunpowder. Any dust of combustible material, even that of cast iron, explodes or burns up with great force. To prove this, let anyone hold a candle beneath a girder or beam in a wood shop and sweep off the fine dust from its top so as to fall on the light, and they will be convinced of its explosive nature. Such explosions are no doubt the origin of nearly all the fires that have been attributed to pneumatic apparatus; as soon as caught, the fire was by means of the wooden pipes immediately carried throughout the whole building, or as far as the air currents extended. So that a whole mill would be fired at once and its destruction certain.

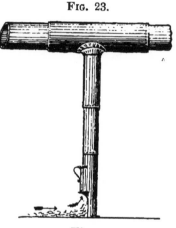

Fig. 23.

Floor.

### PRECAUTIONS AGAINST FIRE.

Besides what has been said upon the danger of wood dust in the last article, a word may be added in respect to other sources of fire, an evil that wood manufacturers have particularly to contend with. Insurance rates for wood-working shops are commonly from three to five times as much as for machine shops and other places, where, if the former were carefully managed, the risk would be no greater. Everyone who has charge of a wood-working shop should continually study the *possible sources of fire.* As accidents do not often happen when they are expected, so fires do not come from sources that are foreseen. Fires are generally mysterious, we rarely know just how they occur, yet there is no want of sources, and considering the little care exercised in most works to guard against fire, the only wonder is they do not all burn down. There is no desire to exaggerate this matter, but to state it in a positive way. The sources of fire about wood-working shops are generally, bearings, smoking, matches, stoves, sparks from the furnace, lightning, and incendiarism, and also the want of means to put out incipient fires, for such want is certainly to be set down among the causes of destructive fires. To consider these several sources:—bearings need not be made so as to take fire; there should be no wood about them, no accumulation of shavings, or of oil and sawdust; smoking, we need hardly say, should not be allowed on the premises; matches are not very dangerous and can be carefully used; stoves are not often required in works where there is steam power, and when they are used, can be made comparatively safe by setting them on an elevated iron platform; sparks from the furnace can only be a source of danger when there is great negligence in the plan of its construction or in its care; and finally, there is but little danger from any or all of these sources in a clean orderly shop. Disposing of the matter in this way, it may be said that it is quite easy to avoid danger from fire. There are none of the things enumerated but what are easily guarded against if taken in time and fully considered. To understand sources of fire is quite another thing, however, from merely thinking of them and being aware of their existence; they must be attended to thoroughly, promptly, and persistently. It is not an easy thing to fire a shop when there is no accumulation of shavings, and a hard thing to guard against fire when there is such accumulation. The floors should be kept clean, no matter what it costs to keep them so, and if the business will not otherwise afford it, the insurance policy had better be paid to a porter to sweep up and watch for fire. The chances are that more will be saved than by insuring. On every floor and in each room there should be kept in some convenient place a number of wooden pails filled with water, not to be used to fill up the grindstone troughs, nor to wash up with, but marked "Fire," and to be let alone unless required for that purpose. It is but little trouble to keep them filled, and a few drops of carbolic acid will keep the water pure in the summer during hot weather. Fifty pails of this kind, that will cost fifteen dollars, are worth more in a wood shop than a dozen chemical annihilators, steam pumps, or other contrivances which men cannot use when excited. A watchman, no matter how stupid he may be, understands a pail of water and will not fail to use it, but would not under excitement be able even to turn a stop-cock, or sound an alarm signal if a fire should occur. The

E

responsibility of these precautions against fire rests mainly with the managers and operators, proprietors do not always understand them, and if they did, cannot watch them. We would therefore urge a carefulness about fires, a thorough study of all that may originate them, and the surest means of arresting them, as one of the first and highest qualifications of a competent manager.

### SPEED OF WOOD MACHINES.

The speed at which machines should run to give the best result, is a problem that operators should understand. To prove that it is an intricate, or at least an undetermined matter, we need only refer to the diversity of opinion among mechanics, and the want of any opinion at all with a great many.

If the speed of a machine could be calculated from that required for the cutting edges alone, we should have a general rule to apply, but the limit of speed is more frequently taken from the conditions of the spindles and bearings, than from the cutting action. Cutter-heads more than 4 inches diameter can generally be moved as fast as the edges require to run to give a good result, say within 5000 revolutions a minute, or 5000 feet of movement with the edges; but when the cutter-heads are smaller, the spindles are not diminished in the same ratio, and the speed must be slower. The cutter movement should as far as possible be a basis for estimating speed, instead of the number of revolutions made by a spindle. A cutter on a 3-inch head, making 4000 revolutions a minute, is only moving as fast as one on a 6-inch head at 2000 revolutions; yet it is quite common, and a habit hard to avoid, to consider all spindles as wanting a common speed of from 3000 to 5000 revolutions a minute, without considering the movement of edges.

Perhaps as good a rule as can be used is to assume a 4-inch cutter-head to make 4000 revolutions a minute, as a base or unit of speed; this makes approximately 4000 feet a minute of cutting movement; then to increase 500 feet a minute for each inch of diameter added to the cutter-head; this makes, at 10 inches diameter, a speed of 7000 feet a minute, and for 16 inches diameter 10,000 feet a minute, which could then become a constant for all larger diameters. This, it must be remembered, is assumed for strong cutter-heads of forged or malleable iron, steel, or brass, and not cast iron, which should not be used for high speeds.

Reversing this rule, from 4 inches diameter, with 4000 feet of cutting movement; deduct 750 feet of the movement for each inch of diameter the heads are reduced; this at one inch, brings the cutting speed to 1750 feet a minute with 7000 revolutions of the spindle, a practical limit. From this we have the Table, page 52, which can be used for reference.

The speed of line shafting should in all cases be as great as the bearings will stand with safety; 200 to 250 revolutions for 3-inch shafts, and 250 to 300 revolutions a minute for 2½-inch shafts, make a good rule, to be modified of course by the kind of bearings used. Countershafts can run three times as fast. 36-inch pulleys, on a line shaft, with 12-inch fast and loose pulleys on the countershafts, is a good arrangement for such shafts as drive cutter spindles.

Machines should, as far as possible, be arranged to start from line-shaft pulleys of a uniform diameter so that they

## SPEED OF WOOD MACHINES.

| Diameter of Cutter-head. | Feet of Cutting Movement a minute. | Approximate Number of Revolutions a minute. | Average Speed of Bearing Surfaces a minute, in feet. | Ratio of Movement in the Bearings. |
|---|---|---|---|---|
| inches. | | | | |
| 1 | 1,750 | 7000 | 875 | 8 |
| 2 | 2,500 | 5000 | 937 | 9 |
| 3 | 3,250 | 4333 | 1083 | 10 |
| 4 | 4,000 | 4000 | 1125 | 11 |
| 5 | 4,500 | 3600 | 1125 | 11 |
| 6 | 5,000 | 3333 | 1145 | 11 |
| 7 | 5,500 | 3142 | 1277 | 13 |
| 8 | 6,000 | 3000 | 1406 | 14 |
| 9 | 6,500 | 2880 | 1444 | 14 |
| 10 | 7,000 | 2880 | 1445 | 14 |
| 11 | 7,500 | 2706 | 1450 | 14 |
| 12 | 8,000 | 2666 | 1465 | 15 |
| 13 | 8,500 | 2615 | 1525 | 15 |
| 14 | 9,000 | 2576 | 1541 | 15 |
| 15 | 9,500 | 2533 | 1551 | 15 |
| 16 | 10,000 | 2500 | 1512 | 15 |
| 17 | 10,000 | 2352 | 1470 | 15 |
| 18 | 10,000 | 2222 | 1417 | 14 |
| 19 | 10,000 | 2105 | 1382 | 14 |
| 20 | 10,000 | 2000 | 1370 | 14 |
| 24 | 10,000 | 1666 | 1250 | 13 |
| 30 | 10,000 | 1333 | 1083 | 11 |
| 36 | 10,000 | 1111 | 987 | 10 |
| 40 | 10,000 | 1000 | 1000 | 10 |

NOTE.—These figures, except the size of the cutter-heads, are approximate only, to give whole numbers.

can be changed, or moved from one place to another, without taking down the line shaft each time to put on a new pulley. There is something strange in the fact that machine makers pay no attention to this matter; even machine tools that have nearly a constant velocity, and require nearly a constant amount of power, are arranged to be driven with pulleys varying from 6 to 24 inches diameter. Most makers, however, are willing to modify their countershafts to suit speeds and pulleys, if a special order is given, so that the fault rests mainly with those who purchase machines.

The cylinders of planing machines being strong and safe, and the rate of feed required as much as possible, they can be run at a speed one-fourth more than that given in the Table.

Boring machines to operate screw-bits should run from 1000 to 2000 revolutions a minute, according to the kind of wood or the size of the bits used.

For all reciprocating machines there is a general rule that applies, which is to run them as fast as they will stand; or, in other words, their work always requires more speed than it is possible to have. This is certainly not a very comprehensive rule, but another rule, infinitely better, is to " use them only when they cannot be avoided," no matter to what purpose they are directed. For ordinary reciprocating machines the following list of speeds is given, for which we trust the reader will not require any special data, but accept it on faith and as a matter of experience :—

|  | Revolutions a minute. | | |
|---|---|---|---|
| Re-sawing machines with one saw .. .. | 250 | to | 300 |
| Scroll saw with sash .. .. .. .. .. | 300 | „ | 400 |
| Jig saws with spring tension .. .. .. | 500 | „ | 800 |
| '„ unstrained saws .. .. .. | 800 | „ | 1500 |
| Mortising machines with movable table .. | 300 | „ | 450 |
| „ „ chisel feed .. .. | 250 | „ | 350 |
| „ heavy, for car work .. | 200 | „ | 300 |

Circular saws can be driven at a speed of 7000 to 10,000 feet a minute. The manner in which they are hammered has much to do with the speed at which they may run, and often when a saw becomes limber and deviates it is a fault of the hammering instead of the speed. When slack on the periphery they will not stand speed, and become weaker and bend more readily when in motion than when still; on the contrary, if properly hammered a little " tight," as it is termed, on the periphery, they become more rigid when in motion up to a certain limit. The cause of this is that steel is elastic, and is stretched by the centrifugal strain in proportion to the speed, which is greatest at the teeth and diminishes to the centre.

If saws have a tendency to spring and a want of rigidity, it can be remedied in most cases by hammering. Cutting wood is like cutting iron; hard wood cannot be cut at so high a speed as soft wood. Any one who has had experience in working boxwood, cocoa, rosewood, or lignum vitæ, will have noticed that a high speed soon destroys edges by overheating, especially with boring tools, or turning tools that act continuously. The use of these hard varieties of wood is, however, so exceptional that the matter need not be discussed here, further than to say that a moulding or a planing machine that is to run mainly upon walnut, ash, oak, or other hard wood, will give a better result if run a fourth slower than for soft wood.

## POWER TO DRIVE MACHINES.

Assuming rules for bands is much the same thing as establishing estimates for the power required to drive machines, and it would be the same in most cases, but not for wood machines. The high speed diminishes pulley contact, and the dust and shavings keep the bands dry, diminishing their tractive force; besides, they must be loose, to prevent the bearings from heating. Experience has demonstrated certain widths as sufficient, and appended is a list of machines with an estimate of the power required to drive them. To determine the size of an engine to drive wood machines, 3 inches of piston area to each horse-power will be found sufficient.

### POWER NEEDED TO DRIVE MACHINES.

|  | No. of H.P. |
|---|---|
| 30-inch surfacing planing machine, one side .. .. .. .. | 8 |
| 30 „ „ „ two sides .. .. .. | 10 |
| 24 „ „ „ one side .. .. .. | 6 |
| 24 „ „ „ two sides .. .. .. | 8 |
| 14 „ planing and matching machine .. .. .. .. .. | 6 |
| 14 „ „ „ with bottom cylinder | 7 |
| 8 „ moulding machine, four sides .. .. .. .. .. | 6 |
| 6 „ „ „ .. .. .. .. .. | 3 |
| 4 „ sash moulding machine, three sides .. .. .. .. | 2 |
| Circular saws for each inch of diameter above the table .. | 1 |
| Mortising machine for light work to $\frac{3}{4}$ inch .. .. .. .. | 1½ |
| „ „ heavy work to 2 inches .. .. .. | 3 |
| Rotary mortising machine, for chair work .. .. .. .. | 1 |
| „ „ „ framing .. .. .. .. .. | 3 |
| Tenoning machine for joiner and cabinet work .. .. .. | 2 |
| „ „ framing .. .. .. .. .. .. | 4 |
| Jig saw for fret work .. .. .. .. .. .. .. .. | 1 |
| Band saw to 1-inch blades .. .. .. .. .. .. .. | 1 |
| Shaping machine, two spindles .. .. .. .. .. .. | 2 |
| Wood-turning lathe .. .. .. .. .. .. .. .. .. | 1 |
| Boring machines .. .. .. .. .. .. .. .. | 1 to 2 |

For grindstones, emery wheels, buffing wheels, hoisting machines, and other details, add one horse-power for each ten men employed; the resistance of shafting, when of unusual length, must also be taken into account.

The power required to operate machines is generally as the amount of material passed through them, so the aggregate must be based upon the length of time, or the constancy with which the machines are run. There must, of course, be enough power provided to drive all the machines at one time, and to their fullest capacity, but in making estimates for rented power where it is employed at intervals, or when but a part of the machines run at one time, the amount used is quite different from what the Table would indicate.

The power required and the power consumed in wood shops are two quite different things. The old saying that time is money, is equally and more obviously true if rendered, power is money. It is an element of cost, just like wages, tools, or material. Power is, however, a less tangible thing, and because it is not seen and handled, is too often allowed to waste and escape under the notice of those who are rigidly careful in other matters. It is common in going into a shop to hear bands screeching on the pulleys, or running half on fast pulleys not in motion, or sometimes a machine is blocked to keep it from starting, with the bands dragging on the pulleys. All this means waste of power and waste of money, not by loss of power alone, but by the destruction of bands. If a band is allowed to rub on a fast pulley, or any other fixed object, it is at once heated and stretched, and, as it stretches on one side, the tendency is to draw it more on the obstruction; if on the edges of tight pulleys, which is most common, its driving power is impaired to the extent that it is rubbed or stretched on its edges. Whenever a heated bearing is suspected, the rule is to hunt it up at once and correct it; the same thing should be done with the screeching of bands. A band always runs to the nearest end of a shaft, towards the line *a*, Fig. 24, which is the opposite way from what is generally supposed. The old theory that a band always runs to the highest part may be true, and is undoubtedly true with respect to the convexity of the face of pulleys, but does not apply to pulleys set diagonally to the line of the band. In Fig. 24 it is easy to see that the pulley 1, standing in the position shown, will wind the band spirally, like the thread of a screw, whose pitch is equal to the space seen at 2,

FIG. 24.

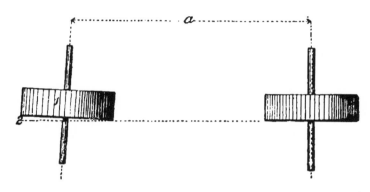

between the dotted line and the edge of the pulley, or, in other words, as the pulley is out of line.

## STOPPING AND STARTING MACHINES.

The resistance offered by a machine in starting, is as the inertia of the parts before they are in motion, or as their momentum after they are in motion. Momentum is as weight multiplied into velocity, hence wood machines, by reason of their great speed, are heavy to start; especially planing and moulding machines with heavy cutterheads. Shifting pulleys, or fast and loose pulleys as they are generally called, are used almost exclusively, and are no doubt the best means for stopping and starting, except idle tension pulleys, which can be used only in particular cases. We should perhaps also except the plan of using an independent shaft, shown Fig. 25, in which 1 is a countershaft, and 2 an idle shaft carrying the stopping pulley. This, although a good device, is difficult to erect and keep in line, besides being too expensive to come into general use. Its merits, aside from these objections, will at once be conceded. In a large mill in Cincinnati, Ohio, the shifting pulleys are all arranged on this plan,

Fig. 25.

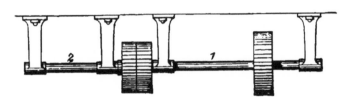

and it is claimed that the extra expense of first cost is more than made up by avoiding the detention incident to having the pulleys loose.

Fast and loose pulleys do very well at low speeds when the shafts are not larger than 2 inches in diameter, and the motion is not more than 500 revolutions a minute, but at the high speeds which are necessary with wood machines, they are often a source of trouble and annoyance. They should be made with great care, and carefully watched for a time when first started.

The holes should be bored and reamed to standard sizes, so that a pulley may be exchanged from one shaft to another, or replaced at any time without the trouble of making a special fit.

The fit should be loose, not too loose, but so as to be felt in shaking the pulley; the hole will show on its sides, from the rubbing of the mandril used in turning, whether it is true or not. A little time spent in looking after these things before starting, often saves detention and accident afterwards, and as the operator has the care, and generally the responsibility of loose pulleys cutting, it is important that he should understand the cause of the difficulty and how to correct it.

At the risk of recommending a plan that seems to be theoretically incorrect, it is suggested that for high-speed

Fig. 26.    Fig. 27.

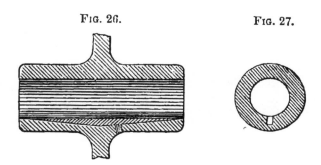

loose pulleys, there should be an oil groove cut, as shown in Fig. 26—a deep narrow groove parallel to the shaft, and tapering from the ends to the middle, as shown in the sections, Figs. 26, 27. Such grooves would be supposed to cause an unequal wear in the hole because of the surface cut away at one side, but it will not be found so in practice.

Fig. 28.

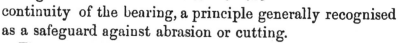

A better, although more expensive plan, is to have grooves cut through the hub, as in Fig. 28; these can be filled with antifriction metal, or wood. The grooves break what is termed the continuity of the bearing, a principle generally recognised as a safeguard against abrasion or cutting.

The proportion of the hubs has much to do with the performance of loose pulleys. A common custom is to make the hubs of fast and loose pulleys, of equal length, losing thereby a large amount of bearing surface that may with advantage be added to the loose pulley, and is not required for the fast one. Fig. 29 is the proper plan of arranging such pulleys, especially for wood machinery, where high speed and wood dust are to be contended with.

FIG. 29.

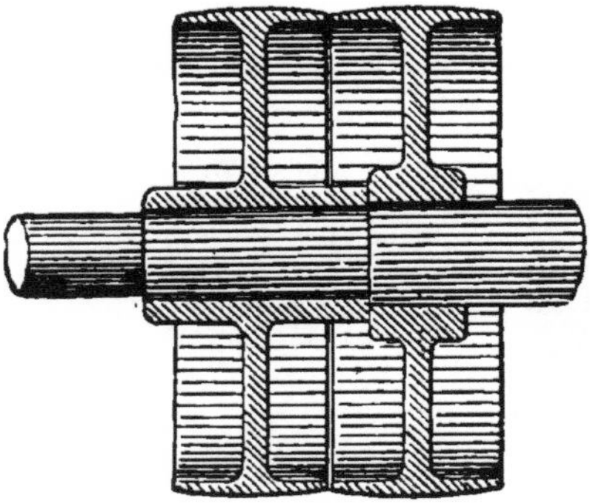

Loose pulleys running on studs or fixed shafts cannot be oiled by means of oil holes drilled in the hub; when a shaft is in motion and the pulley is stopped the oil is drawn in rapidly, but when both are still the case is quite different; in such cases the oil-ways should be made in the shaft or stud instead of in the pulley. This applies in all cases where gear wheels or pulleys run loose on a fixed axis.

Idle pulleys, or more properly brake pulleys, are perhaps the best means of stopping and starting machines or shafts in any case when position allows their use. Any band running at an angle higher than 45 degrees can, as a rule, be operated by a brake pulley; which is not only a very effectual means of stopping and starting, but has the important advantage of regulating the tension of the band to suit the work, and also increases its lap and power.

Wood shops are especially instanced, because a band at any other than a very high angle cannot be operated in this way unless the surfaces are sufficiently dry and

smooth to allow them to slip on the still pulley. As the
bands of a wood shop are usually in this condition because
of the dust, brake pulleys can be used with advantage in
a great many cases, particularly on the larger bands, and
when the driving pulley is below. This latter case allows
the band to stop with the top pulley; but if the angle is
as much as 60 degrees, Figs. 31 and 32, the driving pulley

FIG. 30.                          FIG. 31.

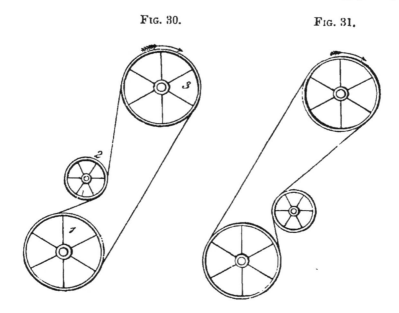

can be above, and the belt will run loosely around the
bottom pulley without injury if it is not too heavy and
there are flanges or guides to keep it on. In Fig. 30, 1 is
the driving pulley, 2 the brake pulley, and 3 the driven
pulley. The brake pulley must always be placed on the
slack side when the bottom pulley is the driver, or as in
Fig. 31, where the upper pulley is the driving one.

Besides the advantages of regulating the tension and
increasing contact, brake pulleys can be used to guide

a band by changing their axes, a very important matter in the case of large driving bands; they also require but one-half the room and width required for shifting pulleys.

---

## ACCIDENTS FROM WOOD MACHINES.

A machine attendant who has not carefully studied the many sources of danger and accident to which he is continually exposed, has neglected something which may cost him a limb or his life at any time. There is always more or less danger from sources that cannot be foreseen, and therefore cannot be provided against, without running risks from dangers that are understood.

Accidents in wood shops occur generally from carelessness, and a failure to correct some irregularity or risk that was well known, such as cuts by saws or other tools in motion, winding bands, bolts or cutters flying off, winding the clothing, and so on. It is rare to find a man who has been engaged for any length of time in operating woodcutting machines who has not lost fingers, or met with other accidents of a more or less serious nature.

There is perhaps less real risk with wood-cutting machinery than many other kinds, if people were equally careful in working with it. One is not apt to go near a train of wheels, or a large band in motion, without a feeling of dread; such things convey a sense of danger; but a small circular saw looks harmless when running, almost as though it could be handled without injury. Unless a high-speed machine makes a great noise it does not seem to convey an impression of danger.

With one exception, circular saws are perhaps the most

dangerous of all. The hands in many varieties of work must of necessity be exposed to injury, and nothing but continual attention and care will prevent accidents. The mind must be kept on the work, and never for a single instant wander away to other matters.

The writer, during an experience where a number of sawyers were under his charge, noticed that a man who was absent-minded was sure to be cut, and that by carefully observing the disposition and peculiarities of the workmen, men could be selected for the saws who ran but little risk. Whenever a man was detected day-dreaming, or engrossed in thought, he was removed from the saws and given work with less risk; the result was, that accidents became rare, although the work was of a dangerous character, consisting mainly of what is termed blocking and cropping, where twelve or more saws were at work.

Accidents in sawing are generally from the hands being jerked to the saw, and from pieces coming over the saw from behind. In the first case the accident generally occurs from a piece suddenly parting in the line of the kerf, either from a split or a hidden cut on the under side that allows the piece to spring forward so quickly that the hands cannot be checked; sometimes by a piece suddenly rolling over towards the saw when a cut is being made on one side. These are cases when a careful sawyer may be cut; but there are a hundred other ways in which accidents may occur, even by people deliberately placing their hands upon a saw without knowing it to be in motion, a circumstance which has often happened.

In cutting short stuff, a sawyer should use a stick for pushing the pieces, placing his left hand to keep them against the fence, and keeping the stick in his right to push them through. A little practice soon makes this a convenient plan, and one that would be generally followed if it were not that in most American saw benches one has not only to push the stuff through but at the same time hold it down to keep it from rising behind the saw, a matter to be noticed farther on. If the stuff has no tendency to rise behind, there is no excuse for placing the hands near enough to a saw to be in danger, no matter what the character of the work.

In sawing from the side of a piece that is liable to roll over, no other precaution can be taken except close attention and an estimate of the danger beforehand. The best rule is to be ready to let go if anything happens, and it may be remarked that in this as in all other cases where accidents may or do happen, people are seldom hurt from a cause that has been previously considered. Pieces coming over the saw is a danger that is more apparent, gives some warning, and is generally dreaded and watched for by a sawyer, especially if he has seen or experienced such accidents. Many who have worked about saws for years do not know the force with which a piece will be thrown from a saw.

If a piece of stuff 10 feet long is taken behind a ripping saw, and the end dropped on the top, so that its whole length will pass over the top, it will attain a velocity equal to that of the periphery of the saw, a fact that is easily proved by examining the marks of the teeth toward the last end, the pitch of which will equal that of the teeth on the saw. An accident of this kind will sometimes happen from a green or wet piece closing on the saw, but in nineteen cases out of twenty the fault is in the gauge or fence, which for some unaccountable reason seems often to be arranged with a special view to causing accidents.

Fig. 32 shows the correct method of arranging gauges for circular saws.

We often see saw benches from 8 to 10 feet long with a saw in the middle, where no one can reach it from the end, and the work being done with the greatest inconvenience; the gauges not only extend past the saw, but are often longer behind than in front.

It is evident that if a gauge extends beyond a saw it cannot be set parallel, but must, in order to free the stuff

FIG. 32.

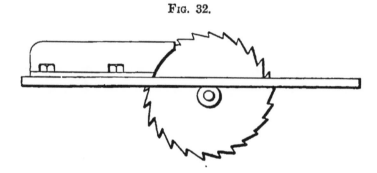

behind, stand at an angle; and the result is that the pieces are lifted behind and thrown over.

Many fatal accidents occur from flying pieces, which from saws of average diameter, usually strike a person in the breast or waist, often causing instant death—sometimes scarcely leaving a scar. Three fatal accidents of this kind happened within as many years to men personally known to the writer. A thick plank hinged so as to hang directly above the saw, heavy enough to stop any piece coming over, makes a safeguard against such accidents, but it hides the rear of the saw from view, and is not needed if other precautions are attended to.

Circular saws were mentioned as second among the

F

dangerous machines of a wood shop. Irregular moulding or shaping machines should be placed first.

Safety shields of various kinds have been devised, most of which protect the hands, but are in the way, and can generally be found hanging on a wall somewhere in the vicinity of the machines. No safety device that impedes or increases labour will ever be used in this or any other case, and the best plan is to carefully consider how accidents may happen and what precautions will prevent them without interfering with the work.

In shaping machines, the danger is from having pieces snatched by the cutters, either by a splinter raising or when the angle of the cutters is such as to cause them to catch, both of which can be in a measure guarded against by having the angle of the edges very obtuse, which generally suits the nature of the work besides promoting safety.

A great share of the work performed on shaping machines, especially such as is extensively duplicated, can be held on forms fitted with clamps as in Fig. 33. This arrangement fully protects the hands, besides making better and faster work.

FIG. 33.

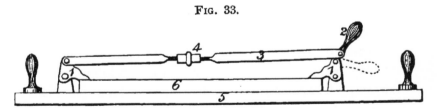

The form shown at Fig. 33 is adapted to shaping chair-stuff, hames, billet frames, or other work, when there are a number of pieces of the same pattern to be operated upon; 5 is the pattern and main frame on which the clamping jaws are mounted, 6 is the piece to be moulded. The jaws 1, 1 are operated by the tension rod 3 and the handle 2, which locks the jaws when thrown down in the position shown by the dotted lines, making a toggle-joint, which is the only safe fastening when there is jar and concussion. The amount of force used in clamping is regulated by the swivel screw at 4, which can also to a limited degree be used to adjust the jaws for pieces of varying thickness.

The edges of the form 5 are shaped to form a pattern and guide. Some pins set in the table top to form fulcrums adds to the convenience and safety of these forms.

The kind of clamps shown are the only ones safe to use. Screws, spurs, or wedges—in fact, anything except a toggle-joint—may give way at any time and lead to accident.

The safety of operating shaping machines depends much upon the form of the cutters; if they have an obtuse angle and stand in a radial position, there is but little tendency to snatch pieces, and the cutting will be effected as easily and much smoother than with cutters standing in an acute position; the angle of cutters will be noticed under another head.

Accidents often occur from winding bands, generally caused by bands becoming fast between pulleys set too near together.

Pulleys on a line shaft, that are separated only an inch or two, are danger traps that may at any time cost a life or lead to destructive accidents. There should always be a space between at least one-third more than the width of the bands, and as much wider as practicable. Bands running too near together are also a source of danger; if one breaks it is apt to be overrun by the other, and both of them wound about the shaft, and as the supports for shafting are often not strong enough to break the bands, the whole is torn down.

F 2

There is always danger in throwing on bands when the pulleys are in motion. It would be of little use to argue against the practice when it will have no influence to prevent it; what is better will be to give such instruction as is possible to lessen danger.

One should not attempt to throw on large bands until practised with small ones at low speeds. There is nothing in a shop learned so blindly as this; no one can, as a rule, tell how to put on a band, or even offer a suggestion, except it be to keep your hands out, or to get on the right side of the pulley. It is learned by accident, as we may say; and yet there is one thing which if understood will save nearly all the experiment, and at the same time the danger, for the danger does not come from throwing on bands so much as the failure to do so. *The hand must move as fast as the pulley;* that is the whole art. By observation it will be seen that the only difference between the skilled and the unskilled rests in this matter of moving the hand with the pulley. One person will throw on a band instantly, apparently without effort, and without a thought of failure; another will try several times, and then, from desperation, attempt to force it on, and burn his hands by friction, or do something worse in the way of accident. As before said, the difference consists in the fact that in successful attempts the hand was moved as fast as the pulley, and in the other case it was not. There are of course other conditions to be observed, but this is the main one.

If the band is long and horizontal, the centre, or bight, as the sailors call it, should be held up, and the slack should be mainly on the " taking-on " side; this provides in a measure for overcoming the inertia.

Large belts, unless very long, should never be thrown

on when pulleys are in motion, but drawn together with clamps and joined. If they have to be thrown on, the pulleys should be stopped and the band lashed to the face of the pulley, the shafts then turned slowly by hand until the pulley has made a half turn, and the belt is on, when the lashing can be removed.

Accidents from winding the clothing are of great frequency in wood shops, but unless from the line shafting, are less serious than in other places. The high speed is a safeguard in such cases. The body cannot be drawn in and revolved about a spindle or shaft that is running at a high speed; the greatest danger is from slow shafts, making from one to two hundred revolutions a minute. Set screws are generally the cause of such accidents.

There should be entered a general protest against all exposed set screws. Many machinists avoid them wherever they can, and in some shops they are not permitted on machines about which men work, and where there is danger; but this is exceptional, and it is common in wood machinery of the present time to find them not only to hold augers and other tools, but in collars on the ends of shafts to keep loose pulleys on. This is unmechanical, and most dangerous as a plan of retaining loose work on a shaft, at a place where bands have to be thrown on and off and oiling done. A nut on the end of the shaft is neater and safer.

Machine attendants often have under their charge unskilled hands, boys who have had no previous experience, and there is great responsibility resting on them in this matter of accidents; a novice uninstructed and uncautioned is liable to meet with accidents that will cost him a finger, a limb, or his life. The dangers of machinery are to him secret traps set for his destruction.

It is not necessary to appeal to the sympathy of skilled men in this matter, for, as a class, wood workmen have but little of that foolish jealousy that in some other trades leaves the young apprentice to learn of danger as he best can.

Accidents from flying cutters, or bolts thrown from cutter-heads in motion, are of rare occurrence. To one who knows nothing of the thing practically, the chances would seem equal, for cutters to fly off or to stay on, when their weight, work, and speed are taken into account. Accidents rarely happen from this cause, however; there is an instinct of danger from cutters that always keeps an attendant on his guard, and anything that flies from a revolving cutter-head always moves *in the plane of rotation*, which it is easy to avoid; this fact is realized, and attendants keep out of this plane when in the vicinity of high-speed spindles.

Cutters are generally held by screws that clamp them to a head or block. These screws have two purposes to serve : to clamp the cutter on the head so firmly that the friction will keep it from being driven endwise; and to hold it against the centrifugal strain. Making due allowance for the tenacity of good bolts, and the strength they are supposed to have in such cases, there is a point of strain at which a screw is ready to break, without adding the further strain of the centrifugal and cutting forces; so the danger is rather in overstraining than in understraining. The inexperienced, generally with a feeling of greater security, will screw down cutters as firmly as they can, and the amount of this strain is usually governed by the length of the screw key.

Cutter-screws and bolts should be made of the very best charcoal iron. Steel is not as safe for such bolts, unless perfectly annealed and soft. It is of course stronger than iron, but it is doubtful if it will stand blows and rough usage so well.

## REPAIRS OF MACHINERY.

A woodwork shop employing twenty or more men should have an engine lathe and forge for performing repairs. The engineer generally has time to work these tools, and will find many things to do on them that would otherwise remain undone or have to be sent to a machine shop.

An engine lathe suitable for general purposes in a wood shop of 16 inches to 20 inches swing, to turn 6 to 8 feet in length, with a gap to receive work to 30 inches diameter, can with the necessary equipment of tools be procured for from 400 dollars to 500 dollars.

The tools and appliances required will be as follows :—

Centre and following rests, furnished with lathe.
One 12″ to 16″ independent jaw chuck.
One set of chuck drills, $\frac{1}{4}$″ to 1″ by eighths, to 2″ by fourths.
One set of twist drills, $\frac{1}{8}$″ to $\frac{3}{4}$″ by $\frac{1}{16}$ths, $\frac{3}{4}$″ to $1\frac{1}{4}$″ by eighths.
A set of V thread taps from $\frac{3}{8}$″ by $\frac{1}{16}$ths to $\frac{3}{4}$″, and by eighths from $\frac{3}{4}$″ to $1\frac{1}{4}$″, with wrenches to turn them.
Two chucks for drills fitted to the lathe.
Six each, 4″, 6″, and 8″, clamp bolts, $\frac{5}{8}$″ diameter.
Lathe drivers from $\frac{1}{4}$″ to 2″ by fourths, from 2″ to 4″ by $\frac{1}{2}$ inch.

Lathe tools as follows :—

Four diamond tools, right and left.
Two side tools, right and left.
Four square tools, $\frac{1}{8}$″, $\frac{3}{16}$″, $\frac{1}{4}$″, and $\frac{3}{8}$″ wide.
Two V tools for threads, one bent and one straight.
One inside thread tool 3″ long.
Three boring tools, 3″, 5″, and 7″ long.
One round end tool,

making 17 in all. These tools should be ordered with the lathe, so they will fit the tool post; and besides have the advantage of being properly made and tempered.

A portable forge from 30 to 36 inches diameter, with a sufficient outfit of tongs, and a cast-iron anvil, will cost from 50 to 60 dollars. If the whole machine shop investment, including the shafting, is valued at 750 dollars, the interest of this at 10 per cent. a year would be 75 dollars. As an investment, this sum will generally be saved in making countershafts, to say nothing of repairing.

With such an outfit, spindles and shafts of all kinds that go on wooden frames can be made; cutters, when of solid steel, can be cut off from the bar, bevelled, drilled, slotted, and tempered. Pulleys within the swing of the lathe can be bored and turned. Nearly all the small items that appear in the expense account for repairs will be saved.

This plan of doing their own machine work is not recommended for small shops; or, rather, it is not recommended as a paying investment, unless the tools can be kept at work a reasonable portion of the time.

A separate room is required for this iron work. To put iron tools into the same room with wood tools is to make a failure of the experiment; small tools are mislaid, the whole covered with dust. Such a room need not add much to the expense, because a place of the kind is required, whether there are iron tools or not, and the little space required for a lathe and forge does not much increase its size. Grindstones, saw-filing vices, oil, and stores, can all be kept in the machine room, and in most cases one man can do the repairing, file saws, grind cutters, and give out stores, besides doing such new machine work as is wanted and the tools will perform.

An engine lathe will perform nearly all the operations of machine fitting, except planing, and even this can be done to some extent on a lathe that has a strong screw and gearing. For drilling there should be a stem pad, like Fig. 34, to fit the sliding head-stock spindle, and a number of wood blocks, of different dimensions, to build up under the work; these blocks should be at hand, to avoid a search for new ones each time they are wanted.

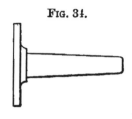

FIG. 34.

In making steel spindles, it is best not to try to anneal them, but cut them off in the lathe by removing the sliding head-stock if the bar is too long, using the chuck and centre rest, which is a better plan than to heat the bars, and will, if we count the squaring up of the ends, be less work than to cut off at the forge. The same rule applies to shafting generally; a bar of any length can be put in a lathe in this manner and cut into pieces as long as the same lathe will turn.

Tempering tools that are not liable to spring is easily learned, and as a wood workman has the advantage of experimenting with edges he may harden, the chances are that with a little practice he can do it better than a smith. Tempering should be learned by every one who uses tools, no matter of what kind. As a process it has little more to do with forging than with any other branch of work, and is a question of judgment rather than skill. Slow regular heating, both before hardening and in drawing or tempering, is the main thing to ensure success. As to the proper shades and degrees of temper, they must be seen to be understood. If a piece of steel is hardened and then polished and reheated on a piece of hot iron, these shades of colour can be learned in one or two experiments. The first shade, pale-straw colour, is right for most wood tools.

## RENEWING SOFT METAL BEARINGS.

Renewing soft metal bearings constitutes a branch of repairing in most all American shops, and while almost any one can make a bearing of some kind, it requires experience and judgment to do it correctly, so the shaft will not be sprung by heat on one side, and the bearing be of proper diameter, with the metal solid and smooth. To this we may add the difficulty of pouring without spilling the metal, burning the hands, or having what is understood as a blow up. In fitting new machines that have moulded bearings, they should be made on mandrils prepared for the purpose, and not on the shafts themselves; but in remoulding them in a wood shop, it is often impossible to have templates for the purpose, because of the various diameters and lengths of the spindles. In such cases the bearings have to be moulded on the shafts or spindles that are to run in them. This operation requires the greatest care to prevent springing the spindles, which will sometimes happen, no matter what precautions may be taken to prevent it. With short bearings, or those that run at a speed of less than 1000 revolutions a minute, there is little difficulty; but in the case of saw mandrils, planing and moulding spindles, shaping spindles, and so on, the bearings will sometimes heat in the most mysterious manner after being renewed.

Whenever it is practicable, both sides of a bearing should be poured at one time and not at two operations as is commonly the case; it requires no more risk or trouble, is sooner done, and with much less risk of springing a shaft. To make them in this manner the shaft or spindle should be first levelled up by placing pieces of brass or wood beneath; the packing should then be fitted, as shown in Fig. 35, with openings to allow the melted metal to run from the top to the bottom, also some vent holes toward the ends to allow the gas and air to escape. This packing can be of paste-board, wood, or of several layers of paper. Soft wood is perhaps the best kind of packing, and is always at hand. After the packing

FIG. 35.

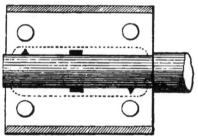

is fitted, the cap can be screwed down firmly and the ends if necessary be luted with clay. If the weather is cold, it is best to heat the cap before putting it on; it will soon communicate its heat to the rest of the bearing and the shaft, which should be turned round so as to be warmed evenly.

In luting the ends with clay they should not be made air-tight; this mistake often leads to a failure. The clay or putty can come to the top of the cap, leaving a free opening or gate for the gas to escape. Bearings that are to be remoulded will, unless burnt out, always contain grease enough to create a quantity of gas when the hot metal is poured in, and unless this gas has free means of escape, the bearing will be blown, and imperfectly filled.

After the bearing has been moulded the gates can be

broken off and the cap loosened by driving it endwise, or by wedging it up with a chisel : the harder kinds of metal are easily separated in this manner, and the softer should not be used for high speeds.

In melting the metal, care should be taken not to overheat it, and to have it at the proper temperature when poured. If it is too hot the shrinkage will be in proportion ; it should be poured at as low a heat as it will run freely. A good plan is to thrust a pinestick into the metal after it melts, and as soon as it will burn the stick or cause it to smoke it is hot enough ; when there are free gates to pour through this test indicates a higher temperature than is required.·

After a bearing has been poured and trimmed, the next thing is to fit it. We are well aware that this proposition will be a new one to many, because such bearings are generally moulded and then started without fitting ; yet there is no risk in asserting that without fitting three out of four will heat at the beginning.

It is evident that if the metal shrinks, as it must do, the bearing will be too small, unless the metal is so firmly fastened in its seat as to prevent it from closing on the shaft. Even if it did not shrink, the bearing would be too close a fit to run cool, so that it must of necessity be fitted. To do this, a round-ended scraper should be used ; this can be made by grinding a half-round file into shape. Those not accustomed to scraping can do better by using the sides instead of the end. The sides of the bearing, which are always too close, should be scraped first; then by putting the spindle in its place, and turning it round, it will mark the spots where it touches ; these can be scraped off until there is a full bearing throughout. The cap can then be fitted in the same manner, and unless a shaft is sprung or otherwise imperfect there will be no heating.

No bearing about wood machines that runs at a high speed, whether it be brass, composition, or iron, can be well fitted without scraping. It would seem that when they are moulded directly on the shaft this would ensure a fit, but a little observation and a practical experiment will prove the contrary.

Bearings that do not run at high speed, for countershafts or line shafting, can be made by winding a layer of paper about the shaft before casting them; it not only provides for the shrinkage and brings the size right, but being a good non-conductor of heat, it prevents the metal from being chilled on the shaft, and will always ensure a sound smooth surface. A sheet of writing paper can be wound around the shaft and tied with a string outside the bearing, or a long strip of paper that is cut parallel and straight can be wound spirally on the bearing and held by the lips at the ends or tied with a cord. There is no fear of having the bearings too large by this plan; it is the opposite that is to be guarded against.

As to the material for moulded bearings, there is no plan so good as to send to a responsible house which prepares alloys and purchase the metal, explaining its purpose and leaving its composition to the manufacturer.

In attempting to mix the metal there is generally more lost by oxidation and other waste than the profit of the regular smelter amounts to; besides the composition is rarely right, and seldom well mixed.

For bearings that run at high speed the best metal is none too good.

We may add on the general subject of the material for bearings in wood machines, in which every wood manufacturer is interested, that moulded bearings made from the best alloys, the metal hammered in and then bored, are no doubt the best of all; what is required is large surfaces, a good fit, and round turning for the spindles.

## LUBRICATING WOOD MACHINERY.

Considering the quantity of oil used in wood-working establishments, its cost and the great difference between its careless and economical use make the subject one worth attention. There can only be a certain quantity of oil utilized, no matter how much is poured on or wasted, and there is little risk in the assertion, that where a pint is required, four pints are wasted. This waste leads to the use of cheap oil to reduce the expense.

Lubricating is, with most kinds of machinery, a question of economy, rather than of efficiency. At slow speeds, except when there is great pressure, almost any kind of oil will do for lubrication; but in the case of high speed, as in wood-cutting machines, their successful operation depends upon efficient lubrication.

It is not proposed to consider the character of lubricants: they are all grease, or ought to be, and their lubricating power, or endurance, is directly as the amount of grease they contain, sometimes as the amount of other matter they do not contain. It is to be regretted that, among the many exhaustive researches that have been made in scientific matters, but little, if anything, has been done to explain and fix standards for lubricating oils. Every manufacturer is annoyed by the persistent visits of the agents of paraffine oil dealers, who have some Latin,

Greek, or Choctaw name for their compounds, which are represented as having some peculiar power of lubricating. The fact is, as before stated, that their worth is as the amount of grease they contain; and as the market value of grease is nearly always constant, the different grades of oil can be considered as representing it in various states of dilution.

Next to the quality of the oil the most important matter is how to apply it economically to bearings.

Constant lubricating is divided into the two methods—circulating the oil in bearings, using it over and over again; and feeding it to the bearing as it is worn out and then allowing it to run off. The first method includes what are termed self-oiling bearings, constructed with cells or oil-chambers beneath the shaft, from which the oil is fed up with wicks, or in some cases through small holes by capillary attraction, and after circulating through the bearing runs back into the oil-cell to be again fed up until worn out. To pour oil on a bearing at intervals, from a can, is to waste three-fourths of all that is used, even if done with ordinary care, and this plan is not to be considered except in cases where no other can be applied; so the choice rests between circulating oil-cells, and oil-feeders.

In the case of self-oiling bearings the wicks are generally inaccessible and out of sight: the arrangement cannot be applied to bearings at pleasure, but must be specially constructed when they are made; and more important than all, the workmen, as a rule, have but little confidence in what they cannot see, and apply oil as often as though there were no oil-cells.

With the glass oil-feeders now used, the oil is fed to the bearing as it is needed; the supply of oil can at all times be seen ; and such feeders can be applied to almost any bearing, no matter what its construction.

There is, however, this objection to the last plan, that the oil will be fed and wasted when the machine, or bearing, is not running, a difficulty not likely to be avoided without adding complication.

This waste is, however, more than compensated in the fact that the workmen have confidence in such feeders, and will take care of and rely upon them, which is not the case with concealed oil-cells.  A prominent engineering firm has by careful experiments determined that a given quantity of oil will last a longer time and give a better result if fed to the bearing from the top and allowed to run off when worn out.

The wicks should be of wire wound round with textile material, ordinary wicking for instance; which can by closing it together or stretching it on the wire be made to feed more or less as required.

All the bearings of wood machines that run at a high speed should have tallow-cups, no matter what other means are employed to lubricate them ; these cost nothing, and are equivalent to placing a sentinel to avoid accident in case the ordinary means of oiling should fail.

FIG. 36.

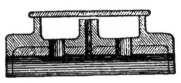 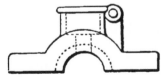

Fig. 36 shows a common box-cap with a tallow-cup as they should be arranged whenever there is room.  Oiling is effected through a centre hole while the cavity around it is to be packed with tallow. If a bearing heat, the tallow is melted, and runs through the holes at each end. These holes should be as large as the size of the shaft will admit, so that the tallow can remain at all times in contact.

Tallow alone is too hard, it requires too much heat to melt it except in warm weather, and should be mixed with lard, when necessary, to give a proper consistency.

For bearings that run at the highest speed a good plan is to cut a narrow groove along the top and bottom, which, if filled with felt, or soft wood, retains and distributes the oil over the surface, and forms a lodging-place for dust or grit that may get into the bearing.

## THE CARE OF BEARINGS.

When a bearing becomes hot, a machine stops; if on the engine or line shafts, all the machines stop; so that it is an important matter to know how to treat such cases. To remove the cause is of course the first thing to be done ; but the cause is sometimes not easy to determine. Aside from becoming dry for want of lubrication, the cause of heating may be want of truth in the shaft, either from not being round or from being sprung; it may be for the want of a fit, and lack of surface, from being too tight, or from too much pressure for the amount of surface.

When a bearing heats, if the shaft is small and can be freed from gearing and bands, the first thing is to see if it is loose enough ; if so, the cap should be screwed down until it binds a little, and the shaft turned by hand, watching carefully whether it binds at one place more than another; the least irregularity can be discerned in this way, and indicates that the journal is not round, and re-

G

quires turning.   If the shaft is crooked, it is detected by
holding a point against it while running—a matter that
any one understands.

If none of these things appear, the shaft should
be taken out to examine the bearing, and see where
the shaft bears, whether at one end only, or on a line
through the bottom, or on the sides.   This want of surface is a common cause of heating with the bearings of
new machinery, and perhaps the most common when bearings have been remoulded.   The remedy is to scrape off
the points where the shaft bears until it touches throughout, as explained previously.   Good oil should be used
in starting, and if necessary the bearing kept cool for a
time with water.

No faith is to be placed in compounds of plumbago,
salt, soap, or anything of the kind ; they may have claims
as lubricants, but it is generally a waste of time to try to
conquer a hot bearing by any other plan than to correct
the mechanical defect, whatever it may be.

## THE PRINCIPLES OF WOOD CUTTING.

It was intended to confine this treatise as much as
possible to practical shop matters, and not to include the
principles of machine construction or of machine action ;
but it is evident that a mechanic qualified to take care of,
to set, arrange, and adjust, or to advise ways and means of
working with cutters, should proceed upon general principles and understand the theory of their action.   Therefore the following brief article on the subject, from the
'Treatise on the Construction and Operation of Wood-
cutting Machines,' is inserted.

" Cutting wood consists of two distinct operations; cross cutting the fibre, and splitting it off parallel to its lamination or grain.

" The two operations are in all cases combined; for to remove the wood both must be performed, and to proceed intelligently about the construction of machines and cutters, this principle must never be lost sight of. The greatest amount of power and the best edges are required to cross cut the fibre. To illustrate by a familiar example :—To cross cut a block 12 inches square requires a considerable amount of effort and time, but a single blow will serve to split it in two, parallel to the fibre.

" This principle exists throughout the whole range of wood cutting with the same general conditions in all cases; a boring auger furnishes another example, different from the one given as an operation, but the same in principle.

" In boring the main power is required to cross cut the fibre with the 'spurs' or 'jaws,' while the wood is split off and raised from the bottom of the hole without much effort; the spurs require frequent sharpening, must have thin edges, and are soon worn away; while the opposite is true of the radial or splitting edges, which may be blunt or dull, and yet work well enough and without much power.

" Another principle to be observed is that the cross cutting or cross severing of the fibre must precede the splitting process; the cross-cutting edges must act first and project beyond the splitting edges. There are no exceptions to this rule, which is from necessity carried out in most cases; yet it is not unfrequent to find tools working on the contrary principle, tearing instead of cutting away the wood.

" In some cases the wood is cross cut at such short intervals or lengths, that no splitting edges are needed, yet the operation is the same. A ripping saw is an example of this kind; each tooth cuts away its shaving, transverse to, or across the fibre, which is split off in the act of cross cutting without requiring separate edges. The cross-cut saw is an example of the same kind, although apparently different: the different shaped teeth that are required arise from the manner in which they are applied. With the ripping or slitting saw the plate is parallel to the fibre, and with the cross-cut saw it is transverse to the fibre; the cutting edges in both cases have nearly the same relation to, and act in the same manner on the fibres or grain of the wood; in short, the difference between cross cutting and ripping saw teeth comes from the rotation being with or across the grain, and not from a difference in the operation of cutting.

"The line of the edge is parallel to the plate in cross cutting, and transverse to the plate in slitting. As before remarked, all operations in wood cutting are the same in principle, and can be resolved into some such simple propositions as follow:—

"First.—Wood cutting consists in two operations or processes; cross cutting and splitting.

"Second.—Tools for wood cutting must have independent edges directed to these two operations, unless the wood is cross cut into short lengths, as in the case of saws.

"Third.—The cross-cutting edges must project beyond those for splitting, and act first, as in grooving and tenoning heads.

"Fourth.—Cross-cutting edges will, if applied at ' an angle to the fibre,' act with less power and be more durable.

"Fifth.—Splitting-edges act best when parallel to the fibre, but ' at an angle to the direction of their movement.'

"Sixth.—Cutters for perforating, or 'end tools,' as we may call them, should be arranged to have their action balanced across the centre whenever practicable, to prevent jar and vibration."

These propositions comprehend the whole system of cutter action, and as all wood manufacture is by cutting, they may also be said to comprehend all that is done in working wood.

We shall not attempt to show their application to planing, moulding, rabbeting, sawing, grooving, shaping and other cutters; the reader can observe this himself, and thus will acquire, if he has not already done so, a general idea of principles, that will guide him in making, setting, and arranging cutters for all kinds of work, without fear of making mistakes and without having to try whether this or that plan will work. It will also furnish a clue to the proper form of saw teeth, shearing knives, and other details, about which there is a great diversity of opinion.

---

## THE ANGLE OF CUTTERS.

The views given on the subject and the examples shown are not based upon theoretical inference so much as upon practical experiment. There are some very obscure conditions connected with the action of wood cutters; if they moved as slowly as metal-cutting tools, we could observe and note the process of their action, but when in motion they are practically invisible, and nothing can be determined except by comparative experiments.

A general object among wood workmen seems to be to get as low or acute an angle for cutters as possible, regardless of the particular uses to which they are applied, and then to prevent slivering, or pulling out the wood, by means of caps. There are, of course, exceptions to this rule, especially with small cutter-heads, as in the case of shaping machines, but exceptions are generally necessary from the form of constructing the cutter-head rather than the result of any plans that have reference to the work. It is generally best not to employ caps on the knives of power machines; they are expensive, inefficient to perform the intended purpose, and are unnecessary if proper attention is paid to the angle of the knives.

Any kind of wood, including boxwood, rosewood, soft wood or green wood of all descriptions can be worked without caps, or chip breakers, as they are sometimes called, by giving the edges a proper angle, and attending to other conditions to be noted.

In planing veneers by hand it has long been demonstrated that the plane iron requires a much higher angle than for other work. It is also known that scraping tools with blunt edges are the only tools that can be used in turning hard woods or ivory; in fact with all hand tools the principle of varying angles adapted to the work seems to be well known and generally applied, but when we come to power tools, planing and moulding machines are made with their knives at a constant angle, usually as acute as possible.

In determining the angle of cutters the following propositions are laid down:—

1st. In cutting clean pine for surfacing, matching, or moulding, the angle of the cutters can be as low as practicable to clear the holding bolts.

2nd. An acute angle requires a thin edge, and a thin edge cannot at the same time be a hard one, nor, for that reason, a sharp one, except in working soft clean timber.

3rd. An edge may be hard, and kept sharp, as the angle is obtuse and the bevel short.

4th. In cutting thin shavings the operation is altogether cross cutting, and a sharp edge is more important than a thin one.

5th. As the angle of cutters becomes more obtuse, the shape of the edge approaches nearer to having the same profile as the work, and the cutters for moulded forms are cheaper and more easily made and kept in order.

FIG. 37.

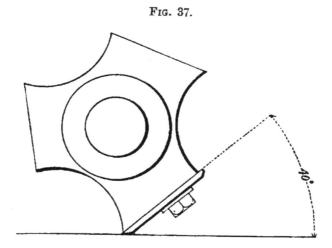

From these propositions we can deduce the following rules, which are recommended to those who have occasion to determine the angle and bevel of wood cutters :—

For planing soft wood, the angle of Fig. 37, of 40 degrees, is suitable.

For mixed work, partly soft and partly hard wood, the angle at Fig. 38 is preferable; it is a mean to comprehend the two kinds of wood.

For working hard wood alone, such as oak, ash, walnut,

cherry, or mahogany, the angle Fig. 39 is best, while for the very hardest varieties, such as boxwood, rosewood,

FIG. 38.

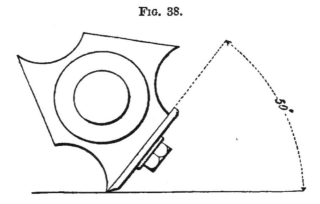

banyan, cocoa, and ebony, working crotch or cross-grained wood, or at an angle against the grain, the cutters should be set as in Fig. 40.

FIG. 39.

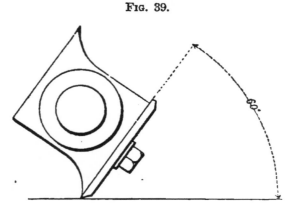

It is becoming of late years a common thing to grind a short bevel on the under side of knives for working hard or cross-grained timber, which is substantially the same thing as changing the angle of the cutters and making the bevel shorter. It is an excellent plan, as it would be impossible to change the cylinders when the machine has a variety of work to do, but by having some extra knives ground at different bevels it becomes an easy

FIG. 40.

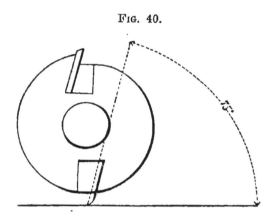

matter to change them, one that will pay well for the trouble, especially if the knives are tempered harder as the bevel becomes more obtuse.

It will be found in practice that a set of knives, hardened to a very pale-straw colour, and with a bevel ground on the face side, just enough to keep the edge from breaking out, will run twice as long, and do smoother work on walnut, ash, or oak wood, and will not pull out the stuff where it is knotty or cross-grained.

It has also become a common practice in some parts of the country to turn the matcher cutters of flooring machines *upside down*, that is, to turn the grinding bevel to the wood; this is an effort in the same direction; a slow change from the necessities of practice, instead of from inference, as it might be. This way of getting an obtuse angle is going a little farther than is recommended here, but to halve the matter by grinding on both sides will be found an advantage in matching hard wood, including yellow pine. The plan is an old one. The Knowles matching heads, introduced about 1850, had this feature fully carried out by having the bevel on the inside of the cutters; they were always considered as being capable of working any kind of timber without tearing, and without clips or pressure pads, yet for some strange reason the plan was not carried out in common practice, probably from being too expensive. We will notice one more fact bearing on this matter, that of machines for making wave mouldings. Such mouldings are cut smooth, and in part at an acute angle against the grain; yet are not torn or spoiled in working; the whole secret of their manufacture, often a matter of curiosity, is nothing more than to set the cutters at right angles to the face of the moulding. The feed movement is given to the wood, and the reciprocating motion to the cutters, which act as scrapers.

## SHARPENING CUTTERS AND SAWS.

Corundum or emery wheels are now generally used for sharpening both saws and cutters. The saving of both time and files, and the more accurate grinding done on cutters, amounts to a great saving when these wheels are properly and fully applied.

Saws, when they can be removed from their mandrils, are now sharpened with wheels, and no doubt could be sharpened with a portable grinding apparatus that could be adjusted to the teeth in such a manner as to sharpen them without removing the saws.

Fd. Arbey, a prominent builder of wood machines, in France, fits his planing machines with grinding wheels that are traversed parallel to the cylinder, and produces with the arrangement, edges that can in no other way be made so true and straight.

Two-thirds at least of the wear of flat or straight cutters come from careless grinding, or over grinding. To grind a cutter up to its edge causes waste. The ground edge is not fit to work with, and it is necessary to whet a new bevel for a working edge. The cutter is then just in the condition it would have been in if the grinding had been stopped short of the edge, leaving a whetting bevel. This is especially true of moulding cutters with an irregular profile, which should, from the nature of their work, if there were no other reasons, have a compound bevel. Fig. 41 shows a cutter with a compound or double bevel, and Fig. 42 one with a single bevel.

FIG. 41.  FIG. 42.

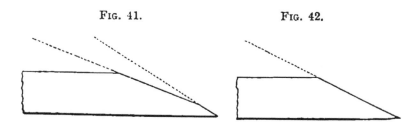

The cutter shown at Fig. 41 is as stiff and strong as the one at Fig. 42, and is more easily whet and ground.

The art of taking care of cutters consists in whetting the edges as the wear requires it, and never grinding quite to the edge, or near enough to weaken it, unless to remove flaws or notches.

For planing-knives, a coarse grain soft stone of the kind known as machine stones is best. It should be not less than 40 inches in diameter, when new, fitted with a tight water box and hood. The stone should be strongly driven, and instead of rubbing for an hour to make an *edge* on a fine hard stone one may in ten minutes finish the knife and have fifty minutes saved to devote to some more agreeable work.

For moulding irons, emery wheels are best. There should, however, for this purpose be not less than five wheels on a spindle, arranged so they can be shifted to different positions, or taken off and put on instantly, as may be required.

The machines manufactured and sold for ordinary grinding purposes are not suited for use in wood shops, and it is better to have them specially made, as in Fig. 43.

The wheels can be moulded on the flanges, as seen in the section, the emery being from

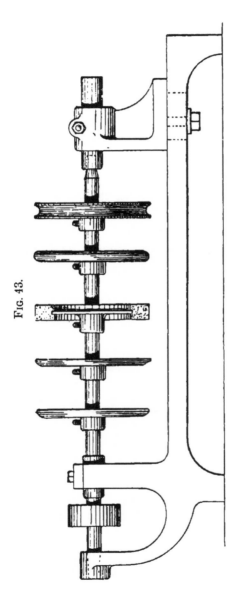

FIG. 43.

2 to 3 inches deep, which is as much as will be worn out in any case; manufacturers of wheels will furnish the disks, or they can be prepared and sent to have the rims renewed on them.

In preparing the disks, or centre plates, there should be two sets, so that one can be sent to have the rims renewed while the others are in use.

Fig. 44 represents a wet-stone machine for grinding moulding irons, used in the large mills in England. It is

FIG. 44.

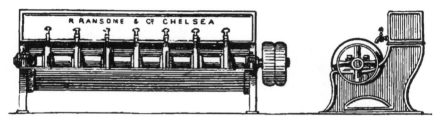

well adapted to the purpose, and with the proper kind of stones they will last a long time and preserve their shape on the periphery. There is no doubt, however, of emery wheels being best, after the men have learned to use them.

For some kinds of planing, thin flexible cutters made from the best sheet cast steel, from 14 to 12 gauge in thickness, will be found a cheap and effective kind of knives; they are now regularly made to any pattern by saw makers and tempered to a hard filing temper, so they can be sharpened without taking them off. To hold them there should be used a stiff steel cap, $\frac{5}{16}$ to $\frac{3}{8}$ inch thick, slightly concave on its under side, and made without having the bolt holes slotted. In many cases thin knives of this kind are used by placing old cutters on the back, instead of having proper caps made, a plan that is apt to lead to a bad result. The successful working of these thin cutters depends upon their being held firmly, and in any case where they have failed to work satisfactorily, it will generally be found that the fault was in the caps, or the bad quality of the steel. Sheet cast steel from the best makers is by no means an inferior article for such cutters, if carefully worked and not overheated in tempering. What will answer for a saw will not do for cutters that have sharp edges, not that a saw is not better if made from fine steel, but the edges are more obtuse and not so liable to break.

These thin knives were patented first by Godeau in France, subsequently by Gedge in England, and perhaps several times in America, so the plan is well patented, if that is to be regarded as a recommendation. In sharpening these cutters, fine mill saw files of good quality can be used. As a rule it is an expensive plan to sharpen tempered steel tools with files, but in this case the material is so thin, and so little to file away, that when the time of taking off and resetting is considered, there is a great saving of cost by sharpening with files. The edge must of course be finished with a stone to make it smooth.

As remarked in the case of grinding planing knives, much time is lost in most shops by using fine grit stones.

The grindstone should be considered an implement for cutting away metal, not making edges, and unless viewed in that light there will be a waste of time and a waste of tools. For large cutters a coarse stone 4 feet in diameter, driven by a band 6 to 8 inches wide, on a pulley 2 feet in diameter is not too much.

For sharpening small tools, such as auger bits, mortise chisels, or others that have angular corners, a dozen good files with various sections, triangular, square, round, half-round, knife edge, flat, and so on, should be provided. Each to have its own handle and the whole fitted into a neat case ; in the same case should be kept several slips of Washita stone, ground to various forms to finish with.

Wood workmen having every facility to prepare lockers and cases generally verify the old proverb by being without them. In machine shops there are, as a rule, places to keep tools and stores, but in wood shops the tools generally lie around loose, and are found when wanted after a good hunt, provided they are not gone out in the shavings and into the furnace. In the matter of files alluded to, how much neater and more economical it is to have a case to keep them in, than to have them lying on the benches, to be used for purposes not intended, and spoiled ; one half the number will do if taken care of, and the whole time of hunting for them saved, to say nothing of doing without them when they are most needed.

To go into a wood shop and find a repairing bench containing three or four files with the tips broken off, one handle to be used between them, a monkey wrench without a handle, or without a screw, a lot of nails, old bolts, paint-pots, and other junk piled upon it, at once indicates the character of the establishment. What the manager does generally determines what the men do, and he can be set down as responsible for the whole. We cannot therefore too earnestly recommend order and system in all things, especially in such appliances as relate to tool dressing, an important one if measured by its expenses, all of which go to the wrong side of accounts.

## SAWS AND SAWING MACHINERY.

Saw benches, which constitute so important a part of the machinery in all wood-working establishments in Europe, are in America comparatively unimportant machines, so much so that they have never been brought to much perfection, and contrast strangely with other machines around them.

The reason for this is that in Europe the material at wood-working establishments is used in the form of logs or deals; the hard wood in logs, and the soft wood in deals. The latter being a technical name for pieces 3 × 9 inches, 3 × 12 inches, and so on, so that all stuff of whatever kind is sawn out to order, and usually finished green.

There are, of course, exceptions, stocks of dry wood being kept on hand in some places, but not on the same scale as in America; a " board yard " contains timbers of nearly every form, sawn in the forest mills ready for use, except seasoning and planing.

This difference in the method of providing timber does not wholly account for the greater one between saw benches there and here. In Europe, a saw bench is employed for uses never thought of at this day in America, and is in fact looked upon as a different kind of a machine, and for different purposes; logs 20 feet long and 18 inches square will be sawn on a common saw bench, not a large one, as that necessitates trucks being provided to carry the ends of the log, and a drag rope for feeding. The saws are thin, and are kept up by 'packing,' a thing quite unknown, or at least not practised, in America. Fig. 45 will show the usual method of arranging this packing. There are two chambers, or recesses, one at each side of the saw, into which is passed a lath or strip of wood, and bound with packing of cotton or hemp, and saturated with oil. This keeps the saws cool, oils and supports them.

Sometimes the packing is done by winding thin flat

Fig. 45.

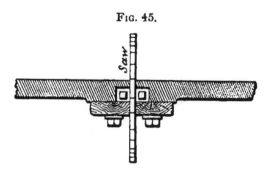

laths of wood, and pressing them in alongside the saws, without having a recess as shown. The saws are besides kept true by this packing ; if there is a tendency to run to either side, the packing is pressed in more close on this side, and the saw forced over into line. Fig. 46 shows a

Fig. 46.

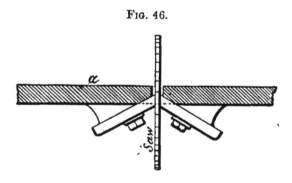

kind of guides that may be employed with much advantage to longer saws, when no better method of guiding and holding is adopted.

The following engravings are made true from elevations

H

of saw benches designed by the writer, in 1879, for the English market. They will no doubt be of interest to American readers.

Fig. 47.

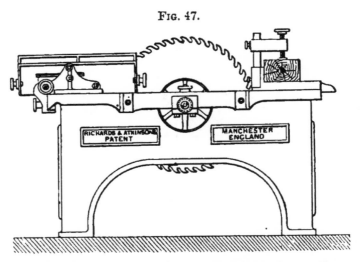

Rear View of 30-inch Saw Bench, with cross-cutting carriage.

Fig. 48.

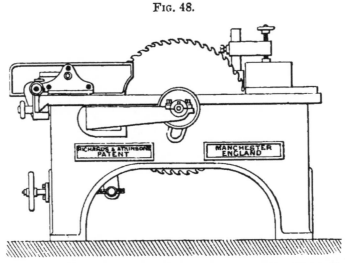

Rear Side of 30-inch Saw Bench, with rising and falling spindle.

FIG. 49.

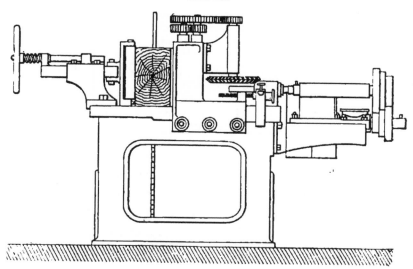

Front View of a 48-inch Saw Bench, with continuous feed.

FIG. 50.

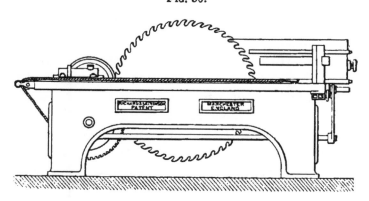

Front Side View of a 48-inch Saw Bench, with drag-feed gearing.

The leading proportions of these benches are given on the following page.

| No. of Saw Bench. | Length of Table. | Width of Table. | Diameter of Largest Saw. | Diameter of Driving Pulleys. | Width of Driving Band. |
|---|---|---|---|---|---|
| | Inches. | Inches. | Inches. | Inches. | Inches. |
| 1 | 48 | 24 | 24 | 10 | 4 |
| 2 | 60 | 30 | 30 | 12 | 4½ |
| 3 | 80 | 36 | 40 | 14 | 5 |
| 4 | 96 | 42 | 48 | 16 | 5½ |

The gauges or fences are made to adjust forward or back, but never to much pass the teeth of the saw. This matter has been noticed in a previous chapter.

In setting saws the custom is to bend the teeth ; some set differently, but bending is the most common practice, so common, indeed, that it is a bold assertion to say that it is wrong, and another plan better. To bend a saw tooth, is not to set it, in a technical sense, and hardly in any other sense, for such set soon comes out in working. A tooth in being set over must have a sharp blow on the inside to stretch the steel, and hold it in position. Saws of any kind can be set with a hammer, quite as fast and much better than by bending the teeth with keys.

For setting circular saws, a frame, as shown in Fig. 52, is convenient. It consists of a rail, 5 × 8 inches, of hard wood, with a sliding block on the top, fitted with wood studs of various sizes to suit the holes in the saws ; on one end is placed a steel laid anvil, to weigh from 10 lbs. to 20 lbs., with its face bevelled off to ten degrees each way from the centre. The saw, being placed on the stud, is moved out or in upon the anvil until the teeth come over the centre ; the anvil is turned until its corner or apex comes across the tooth, in the position shown by the dotted lines, side view, Fig. 51, with the tooth standing over from $\frac{1}{16}$ to

⅛ inch as the amount of set needed and the size of the tooth may require. The tooth is then struck a quick sharp blow with a light hammer, at an angle as shown by the lines *a*, or several blows, if necessary, until the bottom of the tooth is set over as shown in the edge view, Fig. 51.

FIG. 51.

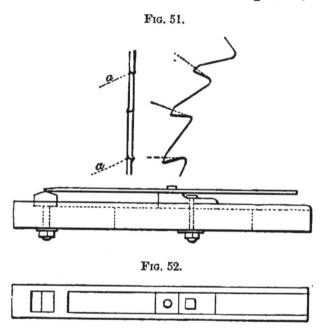

FIG. 52.

This forms a kind of curved scraper edge on the outside, which keeps the side of the tooth clear of the wood, scrapes the surfaces smooth, and will stay there until filed away in sharpening. The teeth will be a little bruised after setting, but this bruising does no harm, and is removed in a single filing. Fig. 52 is a top view of the setting-frame.

All kinds of circular saws and all sizes can be set on the same device. It is cheap to make, always in order, and easily understood. The teeth of cross-cutting saws require setting at a different angle, but can be set in the same manner. Circular saws are too often regarded as a kind of blocking-out machine, to divide stuff into pieces that are afterwards to be brought to dimensions. This, in America, comes from the fact that the great object has hitherto been to save labour, and not, as in Europe, to save material. If a man in sawing has, from the imperfection of his machine, to allow an eighth of an inch on each piece for bad sawing, and his saw cuts out one-third more kerf than is necessary, he wastes his wages, especially with the more valuable kinds of wood.

A saw bench should be carefully and accurately constructed in all respects. Timber can and should be cut to size, leaving only enough to dress it smooth. The frames and tops, more than any other machine, require to be made of iron, so as to withstand rough use, dampness, and wear.

Cutting-off saws are divided into two kinds, those in which the saw is fed to the timber, and those with carriages for moving the stuff, the first for timber that is long and unwieldy, and the second for shorter and lighter work. Carriage cut-off saws are best whenever the material is easier to move than the saw, and the swing or travelling saws are best in the opposite case, a rule easy to remember and easily understood.

Carriage saws have an advantage in their greater simplicity, and consequent durability. The plans of construction are endless, and no suggestions of use can be given here, except that the carriages should be kept square by means of a rack on each end, gearing into pinions on a shaft extending along under the carriage; this admits of its being mounted on rollers, which could not well be used without a squaring shaft.

## BAND SAWS.

What the future of the band saw may be is hard to foretell; but judged upon general principles, such as are common to all saws, there is a probability of its supplanting every other method. Consisting of a thinner blade than can otherwise be used, capable of any degree of tension, and moving at a higher speed than it is possible to run other saws, its advantages are too obvious to warrant any other conclusion. Besides, it cuts square through the wood, and, as a very important advantage, is operated by rotary shafts and wheels running at a moderate speed.

The fear of breaking blades, or the inability to manufacture them, seems to have been for forty years or more what deterred people from using band saws. This trouble has been overcome, and band saws of good quality will do as much cutting as other saws, measured by their value or cost. Joining the blades, from being regarded as the next thing to impossible, has become so simple a matter as to be performed in every shop, and almost by any person.

To first speak of the blades, they should have a high spring temper; if harder, they become more liable to fracture, are difficult to sharpen, and will be broken in setting. A saw that has not a good lively temper is comparatively worthless.

It is quite impossible after a saw is finished to tell whether it is properly tempered throughout; if an inch even of its length has not been tempered, or is drawn by polishing or grinding, it is as bad as though the whole saw was wrong, for such spots cannot be found, and, if they were found, there would be no remedy but to cut them out. People have to trust mainly to the skill and

good faith of the saw makers, and should patronise those
of known skill.

In selecting saws, a good plan to test the temper, if the
saw is not joined, is to roll up the ends, and see if it will
spring back straight or remain bent. If it spring back
nearly to its first shape, the temper is good. The texture
or grain of the steel, which is the only clue to quality,
can be determined by breaking a short piece from the
end of a blade. By unrolling the blade on the floor, it
can be tested as to straightness. The ends, if laid together, will show if it is parallel and of the same width
throughout.

The processes of joining now in use can be divided into
brazing and soldering, the distinction relating mainly to
how the joining is done rather than to any difference in
the processes. In what is termed soldering, the melting
or heating is effected with hot irons, and in brazing the
saw itself is put into the fire.

Brass, spelter, German silver, and other alloys can be
used for joining, any of which make a joint which, if well
made, will be as strong as other parts of the blade, that
is, will stand an equal tension, for the tendency to fracture
is greatest alongside the joints, where the union takes
place between the tempered steel and the portion that is
annealed in making the joint.

For solder joints the silver solder of jewellers is convenient; it is strong, and melts at a low heat. The most
convenient form is to have it rolled in thin strips, so that
pieces the size of the lap can be cut off and laid between.
To make joints of this kind, there is required a strong
heavy pair of wrought-iron tongs and some kind of a frame
to hold the saw straight, leaving the joint free at the ends
to be clasped with the tongs.

Fig. 53 shows a pair of tongs and scarfing frame for soldering the smaller blades.

The saw should be scarfed or tapered at the ends for a

FIG. 53.

length corresponding to one or two teeth, as the pitch may determine. This scarfing must be done true and level, or the joint will not be a close one.

The joint should be cleansed with acid, to remove grease; the solder placed between, and the saw clasped with the tongs, which should have a full red heat. As soon as the solder runs, the tongs should be removed, and a wet sponge or cloth applied to restore the temper in part. The joint can then be filed parallel by using a wire gauge or pair of calipers to determine the thickness, being careful to file the proper amount from each side.

FIG. 54.

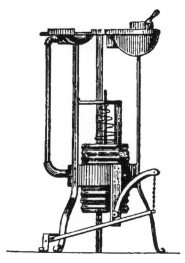

This last is the most difficult part, and requires great care to have the saw parallel and straight, without making it thinner at the joint than at other places.

Fig. 54 shows a forge for brazing band saws, which,

aside from the original cost of the outfit, is the cheapest process, and certainly the best plan of joining narrow blades. The fire is of charcoal, about 2½ inches square ; the degree of heat is accurately regulated by the treadle, which is operated by the foot.

The saw is first scarfed, as in the other case, the joint then wound with brass wire, fluxed with borax, and placed in the fire until the brass melts and runs into the joint ; the saw is then to be quickly removed from the fire, and placed upon a kind of anvil, and the joint pressed together while the brass is in a melted state.

One of the main points in operating band saws is to avoid bending the blades edgewise, which is more easily and frequently done than would be imagined. The wheels require to be so adjusted that the saw will only touch, and not bear against the back guides when not at work ; and as different saws, and different positions of the guides as to height will vary this back thrust, it requires constant attention.

The amount of back pressure is easily determined by placing a piece of wood behind the saw while it is running and pressing it forward, and noting the amount of force required, and then setting the wheels until it bears lightly on the back.

This edge strain, as we call it, is generally provided for by an adjustment of the axis of the top wheel, which every machine should have.

Different forms of teeth, the pitch, angle, and manner of setting, are questions of much importance with large saws that run with power feed ; but for scroll sawing, with narrow blades generally, the matter of teeth has not such importance—a fact that is sufficiently proved by the great diversion of both opinion and practice met with.

The perfection of manufacture and the truth of the blades is apt to be as their width, and beyond 2½ inches wide the steel is not, as a rule, so good, or the saws so true and straight; besides, the tension required for 2½-inch blades is as much as an ordinary machine with shafts 2½ inches diameter will stand.

In respect to sawing logs with band saws, the present limits will not admit of going into the subject. It is a wide one, which does not admit of compendious treatment.

## RESAWING MACHINES.

Resawing, which constitutes a leading branch in the wood shops of other countries, is but a small affair in American mills.

Most planing mills have a resawing machine of some kind, but it is only used for boards too thin to be sawn in the forest mills and safely transported. Timber is nearly all forest sawn, and comes to the manufacturer cut to size, as nearly as can be, allowing for warping, shrinking, and irregularity; not cut first into deals or flitches for transportation, and then sawn again into sizes, as in Europe.

Thin saws and slow feed is the rule for English machines; instead of crowding and forcing one saw to do three times as much as it should, several saws are employed. In America it is common to force a single blade through from 2000 to 3000 feet of boards in a day,—a thing incredible to people who have not seen it, and the result is, as might be expected, bad sawing, and a great waste of both timber and power.

A band saw for resawing American timber should never

exceed 3½ inches wide, nor be less than 40 feet long, the wheels 6 feet or more in diameter; the speed of the saw from 5000 to 8000 feet a minute. The teeth require a coarse pitch, with a deep throat, but of some form to ensure great stiffness, otherwise set cannot be kept in them.

For general resawing purposes, there is no saw better than a compact iron-framed reciprocating machine, to carry from one to ten saws. What may be lost in speed while working but one saw, will be gained when a gang can be used; which would soon be a great share of the time when this system of resawing was once commenced. The blades for such machines need not exceed 14 gauge, and in most cases be thinner.

### JIG SAWS.

What may be said of jig-sawing need not consume much space here. For ordinary wood work a spring-strained fret saw to do the inside, or perforated work, is all that is required.

To set up a jig saw, the strongest place in a building should be selected, over a girder, if on an upper floor; if on a ground floor, there should be masonry or piles set in the earth from three to four feet deep. If the saw is on an upper floor, a counter-balance equal to three-fourths the weight of the reciprocating parts will be best; this throws the vibration on a horizontal plane, in which direction a floor is the strongest of all foundations. If set on an earth foundation no counter-balance should be employed, leaving the vibration to fall vertically, and be resisted by the foundation.

In selecting men to work jig saws, or any saw for irregular lines, two things must be considered—ingenuity and skill to take care of the machine, and the faculty of following lines. Without practical experience, and reasoning from inference alone, one would conclude that almost any person could work a jig saw; but that it requires a peculiar faculty is a well-known fact. A ship caulker, a chipper, or a carpenter, in striking a chisel or in driving nails, cannot tell, or hardly knows, how the blows of the mallet or hammer are directed to the head of the chisels or the nails: in chipping and caulking, the blows are continually varying from one angle to another, apparently without effort or care. The same faculty that guides the hammer and mallet, whatever it may be, is required in jig sawing. The sawyer who has this faculty scarcely knows how he follows the lines; he appears to do so without effort, and depends, in a large degree, upon natural instead of acquired skill. Occasionally men who have great trouble in learning other work make good sawyers; some men cannot learn to turn, others learn with great facility, and a manager who would get the largest amount of work done in the best manner, and in a way most congenial to the men themselves, must watch these peculiarities, as they will be sure to appear among workmen.

Saws for scroll work cut at all angles of the grain, and should have what the nature of the work would suggest, an intermediate form of teeth; not pointed, as for cross cutting, or square, as for slitting, but a mean between, and always in the hook form. A narrow blade is not capable of withstanding back thrust, and should, consequently, be so filed that the tendency will be to lead into the wood instead of crowding back. A triangular file gives a good shape for the teeth of web saws, if they are not too deep, and the pitch not less than one-fourth of an inch. Float files are not so good for filing web saws as the double cut, known as Stubbs' files ; these, although they cost nearly twice as much, are the cheapest in the end, because of the longer time they will last. In selecting web saws, it is best to examine how they have been ground by the saw makers ; if they have been hand scotched, as it is termed, by the grinders, and the bevel is irregular, they will work badly ; machine grinding is the only plan for producing a true blade, when it is narrow, and bevelled back from the teeth.

## PLANING MACHINERY.

After sawing comes planing, and as sawing, except-cutting out, is in America mainly done at the forest mills, planing is the leading operation in most varieties of wood manufacture.

To operate planing machines intelligently and with the best result, one must understand the general principles of their operation, to which we will first call attention.

Under the general name of planing machines are classed, first, carriage machines in which the material is moved in true lines throughout its length by carriages, known as dimension and traversing machines.

Second, machines that reduce the wood to a uniform thickness, or thickness and width at the same time, the stuff being fed continuously between stationary guides and the cutters, such machines being known as surfacing machines, matching machines, and moulding machines.

Third, surface planing machines, that cut away a

constant amount of wood, gauged from the surface that is
planed ; machines that have fixed pressure bars, both in
front and behind the cylinders. The under cylinder of a
double surfacing machine, or bottom cylinders generally,
are examples of surface planing.

These three classes of machines and their operations
are different in principle, and give totally different
results, yet the distinction hardly is recognized or
understood. Every one knows the difference in the machines, and can tell what kind of machine is best for
a certain class of work, but generally from facts gathered
by experience, instead of a comprehensive knowledge of
the principles of wood planing. The different modifica-
tions named will next be noticed in detail.

## CARRIAGE PLANING.

All planing in straight lines has to be performed by
means of carriages, unless pieces have two straight sides
to begin with. A carriage is nothing more than a means
of supplying for the time these two straight sides ; for
when the piece to be planed is fastened to the carriage,
the two are to be considered as one body, guided in two
directions, vertically and horizontally, by the ways beneath.
To make it more plain, we can say that a piece is not
gauged from and by the side opposite to the one being
planed, as in matching or moulding machines, but from
an artificial face, which has been attached, to guide it,
consisting of the platen or table and the guides on which
the table moves. This is the only means of planing true,
and we can hardly expect to see any great change from
present methods of planing out of wind.

The 'Daniels' planing machine,' as it is called in America, invented in 1802, by the celebrated English engineer and mechanician, Bramah, has ever since held its place as the standard machine for planing out of wind. It is no doubt best for some special kinds of work, but is too frequently used; three-fourths of the planing performed by such machines can be as well or better accomplished, and from two to three times as fast, with a cross cylinder. The 'Daniels' planer,' from the nature of things, must be slow in its action; the length of cutting edge that can be brought to act in a given time is the exponent of capacity, and when we consider that the length of edge that can be used is no more than the depth of the cut, the wonder is that they perform so much.

Such a machine with two cutters will, in ordinary work, use only *a half-inch of edge* when taking a cut of one-fourth inch deep; a cross cylinder will, if it has three cutters 20 inches long, represent *five feet of edge*, or 120 times as much as the other machine. The work performed of course is not in this ratio, but the actual cutting capacity is.

The result in working is, that while a 24-inch cylinder may plane 1000 feet of surface without sharpening the cutters, a traverse head will not plane *ten feet* without the edges being equally dull, but as they cut across the wood it can be bruised off with edges that would not cut at all if working parallel to the grain.

The secret of fast planing, we can safely conclude, in not in continuous feed with rollers, which can never make true work, but in increasing the capacity of carriage machines. With a traversing cutter-head the feed is only from 10 to 15 feet a minute; with a cylinder it can be from 40 to 60 feet a minute on a good strong machine. By cutting two sides at once, which is entirely practical in many cases, and presuming that the same time is required in running back, the relative capacity is as one to five in favour of a cylinder.

In the arrangement of a wood-working establishment for purposes which require that a part of the planing be true, and out of wind, there is seldom any absolute need of a traverse planing machine, and unless there is such a need for one, it is best to do without it.

The beating down action of a cylinder, often presented as an argument against their use for thin stuff, is in practice not so serious a matter as it is generally thought to be. A cylinder that has its cutters sharp, and set at a proper angle, will plane almost any kind of stuff without springing it or beating it down.

Both in England and France they manage very well to do all kinds of planing with cross cylinders, not only framing, but flexible stuff, which in America is always planed on roller machines.

There is no question that in the United States too great a share of the planing is on roller machines; the time saved in planing, is generally lost in putting the work together, especially in cabinet work, and similar branches. The tendency to roller-feeding machines is mainly because of their more speedy performance.

---

### PARALLEL PLANERS.

We use the term parallel, because it describes the function of the machines, which is to reduce stuff to a uniform thickness, straightening it in some degree to be sure, but not effectually. Such machines are adapted to but one class of work, stuff that *can be bent or sprung into a straight line,* as it passes through a machine. Keeping this in view, it is easy to determine what work should be done by parallel planing machines. The presumption is that any kind of stuff that will bend in passing through the machine can be afterwards sprung straight in putting it together. Flooring, ceiling, moulding,—in fact, every kind of stuff that is flexible enough, can, and should be, planed on parallel planing machines, which will plane two to four sides at the same time.

A four-side machine, as it is called, although it planes all the sides of a piece, does not do so under the same conditions on each side. Two of the sides are surface planed,—that is, gauged from the surface that runs against the gauges and the bed; the other two are planed parallel, gauged from the opposite side to the one being cut.

The wood is guided by its rough surface before coming in contact with the cutters, and will change the position of its irregularities as it passes through the machine, but will retain them all. By this is meant that a bend in a piece too stiff to be straightened by the rollers and pressure bars, will not be in the same place after planing as before, but advanced to a distance equal to that between the rolls or pressure-bar and the cutters. For this reason, among others, pieces cannot be planed either square or straight on what have been called parallel machines. The top and bottom cylinder will work parallel, and the vertical spindles may work parallel; but as they cannot cut at opposite points at the same time, the piece may change its position between the horizontal and vertical cutters, and be correspondingly out of square. Everyone knows this in practice, and the discussion of it here is not expected to impart any special information as to how the operation may be changed or improved, but to assist in explaining the principles on which different machines operate.

When a piece has two straight sides, and is to be planed all over, or one straight side, and be planed on two, the work can, of course, be sooner and better done on a parallel machine; so that when machines of both kinds are at hand, as is usually the case, pieces can, after planing two sides on the carriage machines, be finished on parallel machines, effecting a saving of time. In furniture making, for instance, if there is a lot of table-tops to prepare, the best side can first be planed on the traverse or carriage machine, and the stuff· be then run through the parallel machines.

Surfacing machines, as they are called, with an endless chain bed, are commonly used for rough work in America, and if properly made in a durable manner, they do very well for the rougher class of work.

The chains should be stronger and better made. The running slats should be chilled on the bottom side, and the fixed bars, or bed, covered with tempered steel. Without this there is no safety in operating these machines, especially on heavy stuff that requires a strong pressure to feed. Surfacing pine-boards gives no test of one of these machines; stiff timber framing, such as car timber, put through one for a few hours will test the endurance of the chain table and its bearing.

In starting a new machine of this kind, great care is required for a day or two at the beginning; the chain and bearing bars have not then come to a fit, and are not smooth and polished. The trays should be frequently oiled. Another fault that is often met with in these machines, is for the chain bed to be narrower than the cylinder and the rated capacity of the machine. This is merely one of those subterfuges too often adopted to convey an erroneous impression of capacity. Traverse machines, for instance, are generally rated as planing the whole diameter of the cutter-head, whereas, as any-one knows, such machines should have their cutter-heads at least one-fourth larger than their rated width.

## SAND-PAPER MACHINES.

Every wood workshop, no matter what the business may be, if the work is to be painted or varnished, can use a set of buffing wheels to advantage. They do not cost much, occupy but little room, and can be operated by unskilled hands when there is nothing else to do.

FIG. 55.

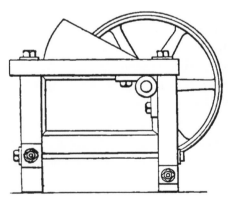

To construct a machine with three wheels, as shown in Figs. 55 and 56, the frames should be from 4 × 6 feet, outside dimensions, the framing from 4 to 5 inches square. Three wheels are better than two, even if but two kinds of paper are required; the two wheels, with the same grade, if laid with the kind of paper used for general purposes, will be worn, as soon as the other, and it will save time. The wheels should be from 30 to 36 inches in dia-

meter, with a face of 8 to 10 inches; they may be made
entirely of wood, but an iron pulley with lagging is best.
The frame should be open on the front, Fig. 56, so as to

FIG. 56.

allow of free access with crooked pieces, and be convenient for the operator. The shaft should be not less
than 2 inches diameter, mounted as shown, to protect the
bearings from the sand as much as possible.

The wheels should be iron, pulleys 30 to 36 inches
diameter with 8 inches face, the rims heavy and turned
true inside and out, with two rows of screw-holes, drilled
¾ inch from the edge, 2 inches apart, to receive 1½-inch
No. 16 wood screws; the holes well countersunk on the
inside. The lag pieces should be either 2 or 4 inches wide,
to match the screw-holes, the joints made carefully, glueing and screwing each one as it is put on. After placing
the wheels in the frame and turning them off true, there
should be a layer of felt, or two layers of thick cloth
fastened on, to form a cushion for the sand-paper. The
outer covering should be a strip of strong canvas two
inches wider than the face of the wheel, and long enough

to go round it, or half round it, as the case may be, the edges notched, as at Fig. 57, so that they will lap over the ends of the pulley, to be fastened with hob-nails. The sand-paper should be in webs or sheets long enough to go round the wheel and be glued to the canvas. The wheels should

FIG. 57.

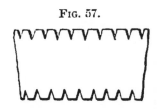

dry thoroughly before being used, and when worn smooth, a new layer can be laid on top of the former one, and this continued until the wheel becomes uneven and irregular, then by drawing the nails that hold the canvas, and cutting the paper across opposite the joint, the whole covering is stripped off, leaving the felt or cloth cushions intact. The canvas can then be placed in water until the sand-paper is soaked off, and again put on the wheel to begin another set of coverings.

It should have been mentioned that the felt covering can be nailed on with small copper tacks, and that in applying the canvas, a strip of paper rubbed with beeswax laid under the joints in the canvas will prevent adhesion from any glue that may go through.

The whole body of the machine frame may be encased to confine the dust, and exhausted by a fan, hoods being placed at the back of the wheels to gather the dust, as seen in Fig. 56.

## HAND-FEEDING MACHINES.

The term shaping, as applied in wood manufacture comprehends all work in irregular lines; a better distinction would be to call all operations shaping, when the stuff is fed by hand. This would include the many improvised plans of doing special work, that cost so little, and save so much, nearly all of which are performed by hand feed.

It is evident that in the great race for automatic machinery, wood manufacturers have gone far beyond the true limit in the use of power feed, and have applied power feed in many cases when the work could be fed to the cutters by hand, and advantages gained both in the quality and cost of the work.

To feed lumber to cutters at a uniform speed, regardless of the state of the edges, the grain of the wood, or knots, is a most unnatural plan, and can only be considered as adapted to the coarser kinds of work; besides, to secure the smoothest and best work, the wood should pass over the top of the cutter-heads, as in hand-feeding machines, and not beneath them. This last proposition would seem to be but a question of relative position between the cutter-head and the wood, but it is something quite different. When material is passed over the cutters, the amount cut away is usually gauged from the side acted upon, and the machine becomes a surface instead of a parallel planing machine.

Hand feed, contrasted with power feed, must not therefore be regarded as meaning two ways of performing the same thing, but as two classes of planing, involving different principles. This distinction is, however, not the most important one between a power-feeding and a hand-feeding machine. The main difference practically is that when arranged with feeding mechanism, a machine is adapted only to some standard kind of work, such as parallel planing, moulding, or grooving, will receive stuff only within certain dimensions, and must be set and adjusted every time the dimensions of the work vary. Besides, in such machines the feed is uniform, regardless of the varying amount that is cut away or the nature of the wood.

A machine that is arranged to be fed by hand is the opposite of all this ; it will receive stuff of any size, will cut away any amount of wood, because the feed can be graduated to suit, and is convertible into a general shaping and jobbing machine, applicable to almost anything within the whole range of wood cutting.

Ten years ago it was most unusual to find a hand-feeding machine in an American wood shop ; whenever the power-feeding machines failed to do what was required, the next resort was hand labour ; but of late years, from experience and necessity, there has been a return to first principles, by the use of hand-feeding machines for jobbing, and they are to be found at this time in most large establishments.*

A singular thing about their use, and one that argues how little the principles of wood cutting are studied, is that such machines have been sold mainly upon trial, and only bought after they had demonstrated their utility. Manufacturers have no confidence in a machine, the merit of which was predicated upon theoretical grounds, and appeared like a discarded thing of the past.

One reason of this is to be found in the common impression that a hand-feeding machine requires a man's time to run it, and that a power-feeding machine does not, a mistake which is easily seen when considered ; in fact, in many cases, hand feed requires no more attention,

---

* Since the above was written in 1873, hand-feeding machines have come into increased use, and in all respects are now filling the conditions pointed out in the present chapter.

and is the faster plan of the two, as bench sawing will serve to illustrate.

Hand-feed machines have been mainly introduced under the name of universal machines, and a common impression exists that their value is due to a combination of several functions, such as planing, boring, and sawing; but a careful investigation of their use will prove their value to be in the adaptability gained by dispensing with the power-feeding mechanism.

A planing, moulding, and general jobbing machine, arranged as in Fig. 58, with an overhung spindle to

FIG. 58.

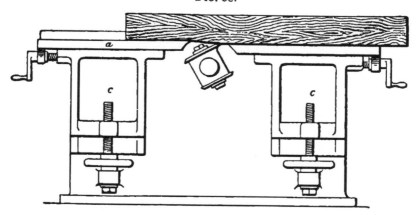

receive various cutter-heads, having a compound table in two parts with independent adjustment, is one of the most useful of hand-feed machines. The tables *a, a* are mounted on movable brackets, *c, c*, which are raised or lowered to suit the diameter of the cutter-heads, and the amount of wood to be cut away. The rear table is adjusted to meet the face after it is planed, and varies from the line of the front one, as the depth of the cut. The figure merely conveys an idea of the general functions of a machine which can be applied to a hundred uses, and will generally have something to do in the way of shaping, moulding, grooving, matching, raising panels, rebating, or other work.

Such a machine corresponds very nearly to the original wood-planing machines; one for moulding and planing very nearly in this form was introduced in America in 1835, but soon gave way to power-feeding improvements, which were capable of performing all required at that day. When modern work demands hand-feed machines it is hard to realize that we must go back to the discarded machines of forty years ago.

## SHAPING MACHINES.

Shaping machines, with two vertical spindles, have now become standard machines in American shops; and we often hear the true remark that they "will do almost anything." When we come to consider why they have such a range of adaptation, it will be found substantially in the principles that have been already pointed out.

This machine, although comprehended in the British patent of Bentham, 1793, and that of Boyd, 1822, was, like many others, a long time in being developed, which only proves that wood-machine improvement is not a question of ingenuity in machine making, but a sequence of improvements in wood conversion.

Machine progress comes mainly from improvements in shop manipulation, and from the wood workmen themselves. This matter is mentioned with a view to directing the attention of operators to processes instead of machines; they must invent plans of performing work, after which it is easy to adapt machines to the purpose. In the case of the jobbing machine alluded to, for instance, if we have the premises or principles to begin with, and know what kind of work can be done in a special manner by a machine, it is then an easy matter to generate the necessary mechanism.

Since the introduction of emery wheels for grinding cutters, the objections to those of solid steel are overcome, and a solid steel cutter, hardened throughout, is sooner ground in this way, than an iron one steel laid. When it is considered that those of solid steel may be one-third thinner and yet as rigid, it becomes an argument in their favour. It is not recommended, however, that the extra thickness be omitted when they are made of solid steel, because shaping cutters are nearly always made too thin. When there is the least spring in them they are liable to break, snatch the piece from the workman, or, what is worse, take his hand into the cutters with it.

When a number of these cutters are needed for shaping machines, and when they are held by angular grooves at the ends in the usual manner, it will be found a good plan to procure several bars of the best cast steel, $\frac{1}{4} \times 1$ in., $\frac{1}{4} \times 1\frac{1}{4}$ in., $\frac{5}{16} \times 1\frac{1}{2}$ in., $\frac{5}{16} \times 1\frac{3}{4}$ in., $\frac{3}{8} \times 2$ in., and $\frac{3}{8} \times 2\frac{1}{2}$ in., in such proportion as the nature of the work may require; cut these bars up into lengths of about 2 feet each, and send them to a machine shop to have their edges jointed and bevelled by a planing machine. This will cost but a trifle, and ensure the uniform width of the cutters, without which no machine can work well, as the spindle is bent to meet any variation of width between the cutters forming pairs.

The cutters can be cut from the bars, shaped and tempered as needed. If there is a very irregular outline

to make, it saves time to drill holes, and break out a part
of the steel in the deepest places, or it can be cut out
at a forge fire without deranging the shape of the cutter,
if care is used. When there is much grinding and
it is to be done on emery wheels, the cutter should be
hardened before being ground, but not tempered until
after being shaped, it will then be clean

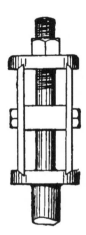

FIG. 59.

and bright, to show the shades of temper.
If solid steel cutters of any depth, say
more than 3½ inches, are used, it is best
to slot them in the centre, and put a
block between with clamping screws, as
in Fig. 59. It may not be needed with
ordinary work, but is always where there
is danger of splinters raising, or pieces
pulling out, that may break solid steel
cutters.

It is safest in shaping, to keep the
material as much as possible between the
person and the cutters. This is the natural
position ; but when fulcrum pins are employed to hold the forms against the cutter-heads, the
operator can in many cases be shielded behind the piece,
or stand exposed as he may choose.

In arranging shaping machines, they should always
drive at as high speed as the spindles and bearings will
stand. The small diameter of the heads requires this, to
attain anything like a standard speed with the cutting
edges; besides, it ensures greater safety to the operator ;
the weight and inertia of a piece will often prevent it
from catching at a high speed, when it would be drawn in
at a slow one. A set of spindles properly fitted should run
at least 4500 revolutions a minute, which with heads 2½

inches diameter gives ¡a cutting movement of less than
3000 feet a minute, much slower than with most other
machines.

The step-bearings for these machines should be as long
and nearly as large in diameter as the top bearings, and
arranged to be flooded with oil. Small tempered steel
points will always give trouble, and have long ago been
abandoned by the best makers for all kinds of machinery.

Have no balance wheels on the spindles, as they only add
useless weight on the steps, which have enough to carry
without them.

### MORTISING.

It was remarked of jig saws that they should only be
used when no other machine could be employed for the
work. It will not be far wrong, and for similar reasons,
to say the same in reference to reciprocating mortising
machines.

In no other country except America have reciprocating
machines been applied to all kinds of mortising, and there
is nothing strange in the reaction we now see going on by
the return to rotary machines for car building and other
heavy work. It is hard to tell which deserves the greater
credit, the ingenuity and care that has kept the reciprocating machines in working order, or the forbearance that
suffers their jar, rattle, and derangement. All reciprocating machines, no matter what their character, if run at
a high speed are open to serious objections—from wear,
breaking, jar, and vibration—but when we add a kind
of duty that consists in heavy blows, like mortising, it
amounts to a culmination of these troubles, and explains

why the "mortiser" in a wood shop is generally out of order and requires more repairing than all the rest of the machines.

As before remarked, it is not our intention to treat of the principles of machine construction further than to give useful hints as to the care and operation of machines, but there is nothing that will teach the care and operation of machines so well as to understand the principles and the general theory of their action. It must also be admitted that as engineers and machinists as a rule know but little of wood-working machines, improvements and changes must be suggested mainly by wood workmen themselves.

We therefore suggest a thorough investigation of this mortising question, to see whether the reciprocating mortising machine has not been applied to many kinds of work which could have been as well or better done by rotary machines. Nearly all mortising in France, and the greater part in England, is performed by rotary machines that cut clean true mortises, without vibration or noise; the question arises, suppose it takes a little longer to cut, a mortise, it is but a small part of the operation in making up work, there are no breakdowns to hinder and derange other things, the work is better done, the tools are not half so expensive, and finally it is worth a great deal to get rid of a reciprocating machine, as a matter of order and comfort in a shop. But even this argument need not be used alone, for some car builders from careful statistics prove that rotary mortising machines effect a saving of time in the end, from the better facilities they afford in presenting and handling long heavy pieces.

There is perhaps no question of the claims of reciprocating machines for light work, and for chisels to $\frac{3}{4}$ in.

wide, or for pieces that are not too heavy to be fed to the chisel. In these machines there is none of the very objectionable mechanism required for a chisel bar feed, and the machines are quite simple throughout. The reciprocating parts can be light and the crank shaft can be placed in the base of the machine, to avoid overhead connections and prevent jar upon a building.

Machines of this kind are suitable for joiner, work, cabinet work, and the lighter kinds of mortising generally, except for chairs ; all other mortising should be done on rotary machines.

In making comparisons between reciprocating and rotary mortising machines, we have to consider—first, the time required to perform the work ; second, the character of the work when done ; third, the skill needed to perform it ; fourth, cost of tools and repairs of machinery, including detention by its derangement; or, briefly, time, quality, skill, and repairs.

To first consider time, it must in the case of reciprocating machines include the cleaning out of mortises after they are beat down, as it is termed, and unless the operator is specially skilled in the proper form of chisels, this cleaning out often equals the mortising. With rotary machines the mortises are clear, but require in most cases squaring at the ends, which can be balanced against the cleaning out in the other case. If a mortise is made in soft wood and without boring, it will be made in less time on a high speed reciprocating machine, but if there has to be a hole bored for starting, the mortise will be soonest made by a rotary machine, which amounts practically to the former proposition that small mortises in light work are soonest made by the reciprocating machines and heavy work by rotary machines.

Speaking of there being but few rotary mortising machines in use in this country, we must except what are generally called chair mortising machines, a kind of rotary one that deserves a more extended use than it has at this time. The rule has been to use these machines on round or crooked stuff which could not be held firm enough to withstand the blows of reciprocating machines; they never fail to do all that is required, and do it well, without much repairing or attention.

The spindles of these machines should stand either horizontally or vertically beneath the work, and run at a speed of 6000 to 7500 revolutions a minute, the vibratory motion may be from 200 to 400 a minute; the cutters or bits should be made from Stubbs' steel, drawn polished rods of the finest grade and used without tempering. The spindles should be bored deep enough to receive from 8 to 12 inches of the rods, so that there will be no waste except the wear, and that the cutter may be set out more or less, as the depth of the mortises may require.

The bits should be held by a conical split thimble fitting into the end of the spindle; set screws are unfit for the purpose; they are often in the way when mortising on angles, and are liable to catch in the clothing.

FIG. 60.

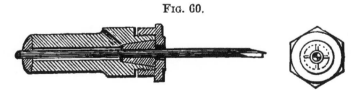

The chuck-end of a spindle is shown in section, Fig. 60, a good device for any kind of rotary tools, where the torsional strain is not too great to be sustained by friction alone.

## TENONING.

Machines for cutting tenons are so well understood, and have been so little changed in a long time, that they are among the most perfect for working wood. Those with a fixed table and a cutting movement of the spindles have come into use for the heavier class of work, especially where tenons are double. With this exception, the American tenoning machines have remained much the same for twenty years past. Improvements have been made in the cutters, the machines have been improved in strength and workmanship, and by the change from wood to iron framing, the manner of adjusting the heads has also been improved and simplified; but for light work an old machine is almost as good as a new one, which can be said in but few other cases. There are some things, notwithstanding these facts, that need improvement. The shoulders of a tenon, for instance, are squared from opposite sides of the piece when it is tenoned at both ends, and it must be bot parallel and straight to bring true work; it amounts to the same thing as using the try square on two different sides of a piece in scribing shoulders, which would not be thought of by a workman. For this we have the remedy of tenoning both ends at the same time, which not only squares the shoulders, but saves time and labour. It also ensures accurate and uniform lengths between shoulders, a matter of no small importance. This plan of tenoning both ends at one operation has gone into practice to some extent in large works in America, and some of the joiners' shops in Sweden and Norway employ the same method. Machines of this kind have been made in England.

An improvement can be made in the carriages. They

K

are generally mounted on slides, and to move them backward and forward is the main labour in operating a tenoning machine; it is not only hard work, but consumes time,
and hinders the operator from holding the stuff, which is
nearly all he can perform with his hands.  The carriages
should in all cases move on rollers, no matter how small
the machine; it is of course more important for heavy
work, and on the larger machines, but in any case it
allows the operator to feel the action of the cutters more
sensitively, and saves time.

Tenoning cutters, with all others that act transversely to
the grain, should be as thin, and stand at an angle as acute
as possible.  The tenons depend for accuracy upon the
edges being straight and true, which requires precision
in grinding and sharpening them, or rather in jointing
them, which should be done when first on the head, and
then a gauge prepared that will indicate the true angle
for the edges: most makers send out such gauges with
their machines, but they nearly always require readjust-
ment by careful experiments.

## WOOD TURNING.

'The Turner's Companion,' with other treatises on the
subject, generally relate to fancy engine-turning for orna-
mentation, and are intended mainly for amateurs, or at
least do not apply to what is needed in a wood-working
establishment.

What is said here will therefore be directed to other
matters that are of more interest to the practical workman,
and while there may not be much said that is new, it will
it is hoped contain suggestions that will be of use.

Turning is an extensive and important branch of wood work, one that has to be performed in nearly all wood shops, and, more or less, on all kinds of work. Every wood workman should learn plain hand-turning; not elaborate pieces, but such things as are met with in general wood work. In joinery, circular work, such as circle top frames, round corners, and columns, have to be turned. In cabinet work, although turning has not so great a share as in former times, it has yet a large part of the whole.

Pattern makers learn to turn from necessity, and the time spent in this way is more than compensated in the aid it gives in learning other work.

The art of turning wood and ivory has always been considered an amusement, and there is nothing in the whole range of industrial processes more fascinating than to shape pieces in a lathe. Some of the pleasure is to be ascribed to the fact that turning is performed without much exertion, and consists rather in directing the tools than propelling them; yet the rapid change of form that is made at will, and the nice skill required in some of the finer varieties of work, make it a most agreeable labour, even to those who are continually engaged at it.

The hand lathe is chief among turning machines. For centuries it was applied to all manner of turning in both wood and iron, without any attempt to guide or direct the cutting tools by mechanism; but of late years, from a turning lathe we have changed to turning "machinery," and so many auxiliaries have been added, that a lathe can now be considered but little more than a device to rotate the piece. In wood turning, all the coarser kinds, and even fine work when there are many pieces of one kind to be made, are machine turned. Nothing connected with wood cutting has been followed more persistently than

K 2

automatic turning, and nothing has met with more failure. A strange fact running through all experiments made thus far is, that they have been successful or unsuccessful as they have corresponded to the action of hand tools. Except in America, but little has been attempted in automatic turning machinery; in the older countries labour is too cheap, and less turning in wood is done.

Hand lathes for wood turning require to be made with more care in some respects than any machine used in wood work; they should run true and steady, as a matter of convenience, and of necessity as well, for no turner can do good work on a bad lathe, or one that is not in order. The cones should be of cherry or mahogany, the wood thoroughly seasoned, and laid up so that the joints will run true. Nothing looks worse, nor more unworkmanlike, than to have the joints in a set of lathe cones to run in a zigzag course; besides, it is just as easy to have them true, by planing up and sawing out the different layers, and then glueing them up on the spindle, using the small cone, which should be of iron, and screwed on the spindle, to clamp them. Iron cones are heavy, cold to the hands in winter, and not liked by practical turners. The cones should be, for a common hand lathe, five in number, rising from 4 to 12 inches diameter. The spindle should have long bearings of hard brass at each end. There is something strange in the fact that while bearings for other spindles are made from three to four diameters in length, lathe bearings are as a rule but half as long.

The shear, or frame, which is seldom furnished with lathes, can be made of wood, which is for some purposes better than if made of iron. An iron shear is cold in winter, generally too narrow on top, and injures the tools, which are sure to come in contact with it. For pattern work, and the heaviest kind of wood turning, an iron shear is for some reasons best, because of keeping the heads in line, and the weight preventing vibration from pieces that are out of balance.

A wooden shear should be made of dry wood, the sides not less than 5 × 10 inches, the top covered with an inch layer of ash or oak, fastened with wood screws, so that it can be taken off and replaced when worn; this preserves the shear frame, and makes a hard surface for the heads to slide upon. Lathe shears should in setting be braced or blocked and bolted, to a wall whenever practicable, especially when there is more than one lathe to stand on a single frame, otherwise one lathe will disturb another in starting rough stuff that is out of balance.

A wood turner needs a good and complete set of tools. It is not pretended that there is anything new in the suggestion, but there never was one more needed; there is no accounting for the want and the imperfection of tools that can be seen with nine out of every ten wood lathes in use. A man may at bench work manage to get along without tools of the best temper, or those properly ground, but no one can turn with satisfaction, or with success, without both, because turning depends upon a sharp keen edge, and in most cases a true bevel, *which forms a rest for the edge of the tools.* The finest steel only will hold an edge, and even then not on all kinds of wood, so that scraping tools have to be resorted to. Except for light work, the scraping tools, cutting off, or square tools, and nearly all except flat chisels and gouges, can be made from bars of steel, and used without wooden handles; if made from ½-inch or ¾-inch square bar, and the sharp corners ground off, they are convenient for pattern turning at least, and much safer than with detachable handles. Tools made in this way should be longer than handled tools ; for pattern turning they may be from 16 to 20 inches long without inconvenience.   For all kinds of light turning, those with handles are of course more convenient.

Tool handles and other fancy articles should be polished in the lathe before taking them out, by first putting on a light coat of linseed oil with a brush, and then using shellac varnish, applied with a woollen rubber, made by doubling heavy cloth to  make two to four thicknesses, and, when doubled, about 3 inches square.   The varnish should be applied to the cloth, then held on the work, pressing hard enough to heat and dry it ; the varnish should be thick, and the operation, to be successful, must be done rapidly.

It may be said that polished work, tool handles for instance, cannot be performed by automatic lathes; such work cannot be made smooth enough to receive the polish, and the polishing if required would have to be at any rate a second and independent process.

No rules of much value can be given to aid a beginner in learning to turn, for turning is an operation consisting almost entirely in hand skill.   One thing, however, may be suggested—that is, cut, instead of scrape.   A beginner at once discovers that his tools will not catch when scraping or dragging on  the wood, and adopts scraping from a sense of danger ; he may at the same time discover that if used in this way the edges of the tools are at once destroyed, but little is accomplished, and the surfaces produced are very rough.

Machine or automatic lathes, as they may be called, consist of four classes :—First, gauge lathes, with a slide rest and tool carriage, after the manner of an engine lathe, for metal working.   Second, lathes with rotary cutting tools, that have a compound motion of the wood and the cutters, both revolving. Third, eccentric lathes for turning elliptical or other irregular forms. Fourth, chuck lathes, hollow mandrils, or rod machines.

The gauge lathe was invented by Bentham, described in his patent of 1793, and has possibly, under some modification, been in use ever since. What is known as the Alcott slide, to be used in connection with an ordinary hand lathe, is but a modification of this machine. The principle of operation consists in a following rest, in front of which is a roughing gouge, to reduce the piece so that it will fit the rest ; behind this rest other tools follow, one to three in number, as the work may require, the rest supporting the piece. The following or finishing tools are generally mounted upon pivoted falls, which slide on patterns, that raise and lower the cutters to give the required shape to a piece. This produces duplicate pieces very rapidly, but if the profile is in any degree irregular the work is too rough for any but the commoner purposes. By tumbling the pieces in a cylinder with leather scrap, after they are thoroughly dried, they can be made smooth enough for painted work, but not to varnish or polish. Gauge lathes have been helped out of this difficulty of making rough work by shearing knives, that come down diagonally behind, and follow the rest, cutting off a light shaving with a thin tangential edge, corresponding to the action of a hand chisel, that leaves the piece true and smooth. This device has been extensively and successfully used, and manufacturers need have no fear in adopting it for any work to which it can be applied.

If a gauge lathe is to be used, it should be a good one. It was a long time being discovered that a gauge lathe for wood turning required to be as accurately, and even more carefully made, than an engine lathe for machine fitting. Such lathes require to be made in the most thorough manner, and will cost a large price from any responsible maker. If the amount and character of the work does not justify the outlay for a first-class machine, it is better to do the work by hand, or with an Alcott slide, than to buy a cheap one.

The spindle bearings of gauge lathes should be made of the hardest brass, set into accurately planed seats, so that they may be adjusted or renewed without trouble. Centres project 6 to 10 inches from the ends of the spindles, have sharp points, and the head and tail points must come together precisely, and keep there—that is, the lathe must keep in line. This must be the test of a gauge lathe, and is one that would condemn nine-tenths of all the engine lathes in use.

Lathes with rotary tools are but little used. The cutters and the wood both running in circles, and cutting inter-mittently, make rough work; it is difficult enough to produce smooth surfaces with either the wood tangential to the cutters, or the cutters tangential to the wood, without having two circles to meet. There has been a limited use of these lathes for turning wheel naves and other coarse work, but nothing to merit a further notice here. We suggest to wood manufacturers that whenever they find this compound rotary motion of both the tools and the piece in a machine to do cylindrical turning, to buy some other; it is a subversion of the true principles of wood cutting, and as such should be employed only when it is unavoidable.

Eccentric lathes for oval turning are among those machines which require special knowledge to manage.

The Blanchard lathe, if driven at its utmost speed, may turn from five to seven hundred small spokes a day.

The best lathes for eccentric turning are those which have the reciprocating movement in the cutter-head, a principle which is followed in all cases except for spoke turning, and in what is called a Handle Lathe. The fact is that no durable and substantial machine can be made that has its spindles and driving gear vibrating on a swing frame. The lathes used for turning gun stocks in the armories are the best in use, and are in all cases made with the spindles to run in fixed supports.

With handle lathes the cutting is generally done by saws which stand the bark better than cutters, and do not spring a piece so much. A cutter-head with six or eight cutters to do the same amount of work as a saw that has from 24 to 32 teeth, must displace four times as much wood at each cut, and the shock and strain upon the piece is nearly in the same proportion. The straighter kinds of handles, such as sledge, pick, hammer, and hatchet handles, can be turned much faster with cutters than with saws, because of the edges being broader, and the feed proportionately faster; but axe handles, or any handle that has short turns or angles, can be best turned with saws. The best plan is to have each lathe supplied with both saws and cutter-heads, so that they can be changed to suit the kind of work being done.

Chuck turning, as it may be called, relates to parallel rods like dowel pins, chair braces, or fence pickets. As machines, such lathes are simple, efficient and labour-saving, cost but little, and should be employed whenever there is anything for them to do. The principle of their operation is the same as the hand gauging tool shown at

Fig. 61, a little device that should be among the tools on every hand lathe.

This gauge tool can be used in turning any kind of parallel stuff, dowel pins, wooden screws, gauge stems, in fact any piece that is in whole or in part straight. Cabinet turning, such as nulling, cottage spindles, or other pieces that are turned straight before being moulded, can

Fig. 61.

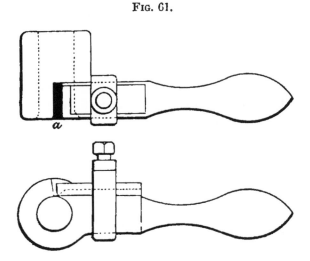

be sized much quicker and more accurately with a gauge tool than with chisels.

They are made of cast iron, are inexpensive, and easy to operate. One stirrup and cutter will do for several sizes by exchanging. The only fitting is to bore them to the size wanted and cut away the throat. In using them the handle runs on the rest, and should be held down firmly; some of the first experiments may be failures, until there is some skill acquired in setting the cutter. The tool will either go perfectly straight, which is its natural and most

easy course, or it will not go at all. Although but little
known, they have been in successful use for years, and are
especially useful in turning the stems of wooden screws,
and other pieces that have to be truly parallel.

## PATENTS ON WOOD MACHINES.

It is thought that, among other things, a short article
on the subject of inventions and patents would not only
be of interest, but probably of use. All engaged in wood
manufacture—proprietors, managers, and workmen—are
at some time either afflicted with a patent mania themselves, or brought in contact with it in others, and the
little that has been written or is known of the history of
wood machinery, together with its recent rapid develop-
ment, has been not only favourable to invention, but also
to deception and mistakes. If anyone before investing
in, or becoming interested in, inventions, or in applying
for patents on wood-working machinery, would look over
the statistics of the past, and see how little has been
derived from invention, or even from the monopoly of
manufacture, by patents on wood machines, he would need
no other caution to deter him from what will, in nine cases
out of ten, result in a loss of time and money. Even in the
case of the few leading patents, on "principles"—we use
the word advisedly—such as Woodworth's patent on planing machines, or Blanchard's patent on eccentric lathes,
but little, if anything, has been gained to the patentees.
The greater share of the revenue has been consumed in
litigation, to defend against infringement, a consequence
that is natural, and will always occur in any attempt to
monopolize the manufacture of a machine after it becomes

valuable. There is something about public sentiment, especially in the United States, that rebels against patent monopoly, and favours attempts to evade patents.

Leaving out the considerations already named, which ought to be quite enough to save at least the greater share of what is each year lost in wood machinery patents, there is one other too often lost sight of—the difficulty and expense of ascertaining the novelty of improvements. The Patent Office, with all the good features of its system, and the examination it gives to cases, does not dare to give any validity to a patent, or to confer a single right that is indefeasible or not conditional; it simply gives the inventor power to prosecute others for infringement, and actual damages on condition of being the true and original inventor, as against everyone else.

What it is intended to notice here is mainly the founding of business schemes with patent monopoly as a base, or constituting a part of the capital. Any failure of a manufacturing business is felt far and wide, both as a loss of capital and an injury to the reputation of the branch of work to which it belongs; and in establishing a business, as in building a house, there is required a good foundation, which in manufacturing should be a demand and market for the product, skill to produce it at as low or a lower cost than others, and capital to do the business upon.

## SUPPLYING MATERIAL.

"A penny saved is a penny earned," is a maxim as old as it is true; and if applied to the purchase of wood to supply to an establishment, it means that anything saved in that way can be added to the profits account.

As to purchasing sawn lumber, it is only a commercial question of quality and value; but other plans of procuring material, without its passing through the general market, are open to some suggestions for those who are within reach of timber.

A great many, in manufacturing articles 'from wood, never think of anything but to purchase sawn timber and recut it, often into small pieces, which could as well be cut from round timber, saving thereby a great share of the cost, and securing better material. As a rule, 200 feet of plank or scantling will cost as much as one cord of timber; a cord of timber, 128 cubic feet, is as a solid equal to something over 1500 feet of sawn stuff board measure; allowing one-half for saw-kerf and waste, it would make when sawn 766 feet of material. A good sawyer, with an efficient machine, will cut up four cords of logs 8 feet long into framing pieces or turning stuff in a day; the waste, after furnishing fuel to drive the saw, is generally worth enough to pay for the sawing, and something over; as one-half is allowed for waste it should certainly make a cord of firewood, worth as much as a cord of round timber.

This would give 766 feet of prepared stuff at the same price that would have been paid for 200 feet of plank or flitches, with the difference that what is cut from round timber would be to true dimensions, while the other would be in planks or boards, and subject to a much greater waste in re-working.

Many of the largest wood manufacturing establishments have already adopted this system, so far as they can, and have continued it successfully for years; it is not expected, of course, that we are giving them information, but a great many never think of it.

Machines for sawing round timber into small pieces not

being generally used or known, it is thought best to give some information in respect to a cheap and useful modification such as is now employed in various works. Figs. 62

FIG. 63.

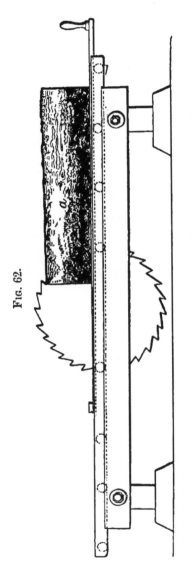

FIG. 62.

and 63 are side and end elevations ; the general dimensions as follows :—

Length of main frame, 14 feet.
Height of main frame, 24 inches.
Length of running board or table, 13 feet.
Length of bearing rails, 16 feet.
Diameter of saw, 36 to 40 inches.
Diameter of mandril, 2¼ inches.
Length of mandril, 42 inches.
Size of pulley, 12 inches diameter, 8 inches face.
Speed of the saw, 1200 revolutions per minute.
Power required, from 10 H.P.

The table is merely a hard wood board, interposed between the timber and the rollers, divided throughout its length but held together by the cross cleats on the ends; the angle irons seen on the end view, at *a*, are used to gauge the stuff; other plans can be used, which may be more convenient ; one is to have swinging gauges fixed to the main frame outside the moving table, so that they will swing round out of the way when the timber is moved: another is to have lines scored on the table, indicating inches or smaller divisions if needed. In sizing stuff that is to be squared after it is cut into deals, a number of pieces can be piled on top of each other and cut at one time to save time and walking. Six cords of timber have been cut into pieces for hoe handles, rake handles, and general turning lumber, in a day, on one of these saws.

Table legs, bedstead and chair stuff, with the greater share of the lumber used in furniture manufacture, can be prepared in this way, either to size, or in pieces to be again sawn after seasoning.

BENCHES FOR WOOD WORK.

Long custom has established certain forms of benches for different kinds of wood work, and while almost any kind of wood work may be done on almost any kind of bench, there are in this, as in most old customs, some good reasons at the bottom. The cabinet maker wants a tail-screw, the carriage maker a standing or high vice, and the pattern maker the back tray, while the carpenter does not care much what his bench may be, so long as it is long and wide.

Since the general introduction of machinery to do the planing, benches have been made higher than when work was done by hand, an improvement that prevents stooping. Thirty-two inches high was once a limit, but now benches

36 inches high are often more convenient than if lower. In any case they should be as high as possible.

The main parts of a bench are the top and vice. For carriage work the vice is the main part; there is, however, no harm in having both as good as can be. The tops should never be made of a whole plank; they are much better if made of scantling, bored at intervals of 12 inches for dowels, and the whole drawn together with $\frac{5}{8}$-inch bolts. One of the bolts can pass through the standing leg of the vice, which should always be gained into and come flush with the top of the bench, and not mortised into the under side, in this way it generally splits the top; besides, the top will not stand the wear opposite the vice-jaw. When a tail-screw is to be used the top cannot well be made throughout of scantling; a wider piece will be needed on the front side, to frame the tail-vice in, but it should be as narrow as possible, and the rest of the top in pieces. Benches should always be large enough; the constant tendency is to have them too small, especially with cabinet makers, who often own their benches, and move them like a tool chest to wherever they are engaged; this has no doubt been a reason for the small size that cabinet makers' benches are generally made. A tray at the back is the common plan, and at the risk of violating the maxim laid down at the beginning about old custom, we must say it is wrong. No one wants to hunt small tools out of a tray; it is never deep enough for the stuff on the bench to clear plane handles, is always full of dirt and shavings, and at best can be considered as nothing more than a plan to save width. If the stuff being worked is wide enough to cover a tray, the tray is of no use; if it is not, there is still no need of the tray, six inches more of width added to the top will be found more convenient for any kind of work, from carving to waggon making. A flush top is easier kept clean and clear, but if it must be divided into a working top and a tool compartment it is better to raise the tool platform *above* the bench, either by laying on a pine board 8 inches wide, or, if the bench is less than 30 inches wide, this board can be set from 5 to 8 inches from the bench, like a shelf, leaving room so that planes will go under it. This will be found more convenient for small tools, and more orderly than a tray. A cabinet bench, to be convenient, should not be less than 30 inches wide, and 8 feet long, the centre of the main vice 20 inches from the end, the top 3½ to 4 inches thick for the whole of its width, for it nearly always costs more to fit up a backboard than the extra wood is worth if the tops are made of uniform thickness throughout : these sizes are nearly twice as large as such benches are usually made. The amount that a man may earn on a bench does not often lead to affluence, even when all conditions are favourable, and to render this amount as large as possible, after skill, the next most important thing is order and good tools, neither of which can be had without bench room.

A vice-jaw for wood work, except waggon and carriage making, should be from 8 to 10 inches wide, 3 to 3½ inches thick, of seasoned hard wood set at a sufficient angle to prevent it from twisting when long pieces are set in vertically. The standing leg should be the same size, and, as before said, gained into the top depth, not full depth, but from 1½ to 2 inches.

Benches for pattern making require to be wider, longer, and higher. A good plan for pattern benches is to make them continuous along one or more sides of the building. The tops need not be more than 3 inches thick : if covered with pine, it prevents bruising the work, and is easier to

L

true up. Such benches should be 32 to 34 inches high, 36 to 40 inches wide; if in sections, at least 10 feet long. The vice should be strong, of the same proportions before given. The screw will be more convenient if of a coarse pitch, so as to act quickly. The square thread screws, coming into use, are well adapted to pattern makers' vices, or for any work that requires much use of the vice. A screw and slide bar should always be arranged so that the vice can be drawn out or closed up to any distance, without helping the bottom along with the hands; this can be done by putting a bearing over the top of the screw inside the nut behind the standing leg, and by having a well-fitting collar key, and also a good running bar at the bottom. It is better to spend a little time in fitting a vice properly, than to stoop or sit down to pull out the vice-jaw at the bottom each time it is changed for different sizes of stuff.

Waggon and carriage makers often use parallel iron vices, which are better for the purpose.

---

### BENCH TOOLS FOR WOOD WORK.

It was remarked in the Introduction that we sometimes see a man without much physical strength, and apparently without exertion, do more work than a strong one who labours harder. No fact is better known to wood workmen than this, but the lesson it teaches is generally neglected, and the matter regarded as a kind of mysterious dispensation, over which there is no control. There is no greater mistake; workmen may have peculiar faculties mentally that enable them to succeed better than others when their work is very diversified and intricate, but so far as bench work is concerned, the difference can be traced mainly to the tools used ; a fast workman generally has plenty of them, kept in order, and in the right place. Hand skill is of course requisite, but hand skill is a result of good tools, and the same spirit that promotes order ensures speed. A man may do good work with poor tools, but such a man takes no pride in his business, and could do proportionally more, and with greater ease, if he had better tools.

It is almost impossible to speak intelligently or specifically about tools without assuming some special kind of work to govern the matter, but as this would exceed the brief limits assigned to the subject here, we will endeavour to treat it in a general way.

The first and leading tools are bench planes, a set of which should consist of one 26-inch jointer $2\frac{3}{4}$-inch iron, one 24-inch jointer $2\frac{2}{3}$-inch iron, one 22-inch fore-plane $2\frac{1}{2}$-inch iron, one jack plane $2\frac{1}{2}$-inch iron, all double irons ; one jack-plane 2-inch single iron, one handle smooth plane $2\frac{1}{4}$-inch iron, one common smooth plane 2-inch iron, one block plane 2-inch single iron—nine planes in all, as a set of bench planes. To these standard planes may be added a panel, plough, and right and left rebate planes. Other planes, such as hollows and rounds, match and moulding planes, are usually shop tools, to be used in common.

For chisels, an outfit should consist of a set of firmer-chisels, from $\frac{1}{8}$th to 2 inches ; two long firmer-chisels for paring, $1\frac{1}{4}$ and $1\frac{3}{4}$ inches ; two socket chisels for heavy work, $\frac{3}{4}$ and $1\frac{1}{2}$ inch ; bench gouges, from $\frac{1}{2}$ to 2 inches ; and a 1-inch blunt scraping chisel. All should be in order, handled, and kept in racks at the back of the bench,

within easy reach. For saws there will be required one
rip and one cross-cut hand saw, one panel saw, one each
12- and 8-inch back saws, with others of a special character,
such as a ramp saw, bow saw, and dovetail saw.

Planes, chisels, and saws, are the main tools in bench
work, and should be of the best quality.

For the convenience of apprentices who desire to select
a set of tools, the following list is appended; it may
contain many more than are required, but will be none
the less useful for reference ;—

> Planes, as before.
> Chisels and gouges, as before.
> Hand, back, and other saws, as before.
> 3-, 5-, and 7-inch try-squares.
> One carpenter's steel square.
> Bench and small hammers.
> One wood mallet.
> One 5-inch hand-axe.
> One 24-inch single-fold slide-rule.
> Oil-stone, slip, and oil-can.
> One pair 4-inch spring dividers.
> One pair 8-inch steel compasses.
> One wooden brace, with full set of bits.
> One set auger-bits, from $\frac{1}{4}$ to 1 inch.
> Two spoke-shaves, $2\frac{1}{2}$ and 3 inches.

To these may be added a number of little things which,
although hardly to be included in a list of bench tools,
will often be wanted, such as a chalk-line and spool, bench
brush, strap block, sand-paper block, wood straight-edges,
plumb-line, spirit level, or bench hooks, which can be
supplied as needed, but should be owned by the workman,
and kept at the bench, each in its proper place.

When a man learns bench work, he should do so

thoroughly. He should study and observe the various modes of performing work which he sees around him, and estimate their advantages. The fact that bench work is mainly done by the piece in wood shops would be, as one would think, a sufficient incentive for workmen to study it carefully, with a view to increase their earnings, but strange to say the facts do not permit such a conclusion.

In using bench planes it is a good plan to learn to plane with one hand as much as possible, especially with jack planes.

To keep both hands on a plane makes one of two things necessary, either to walk along and carry the body with the plane at each stroke ; or else to plane by short strokes, making a kind of chipping operation. A man can stand in one position and plane the length of a piece 4 feet long with one hand, and propel the plane with just as much force, and when he has learned it, with more force than if he used both hands. If a brace pin is used in the side of the bench, he can, in roughing out with a jack plane, do twice as much in a given time as he could by grasping the plane in both hands and moving his body with it. Granting this proposition, which will be fully proved by an experiment, and following it until learned, is it not strange that we rarely see planes used in one hand ?

Another thing connected with dressing up stuff which may save time and labour, is the use of the try-square. Supposing that a piece is being jointed or squared in the vice, the custom in trying is to remove the plane, put the square on the piece with the blade on the top, and then stoop down to look under the blade, generally low enough to bring the eye level with the piece. This can be ·done with half the trouble and with more accuracy

FIG. 64.

by placing the head of the square on the top of the piece, as in Fig. 64, and looking down along the blade at the side. To do this the plane need not be removed from the piece, the body is kept erect, and in the case of a thin board instead of having but its thickness to gauge from, there is the whole length of the square blade.

LONDON: PRINTED BY WILLIAM CLOWES AND SONS, LIMITED,
STAMFORD STREET AND CHARING CROSS.